Nanofluids

Nanofluids provides insight to the mathematical, numerical, and experimental methodologies of the industrial application of nanofluids. It covers the fundamentals and applications of nanofluids in heat and mass transfer.

Thoroughly covering the thermo-physical and optical properties of nanofluids in various operations, the book highlights the necessary parameters for enhancing their performance. It discusses the application of nanofluids in solar panels, car radiators, boiling operations, and CO_2 absorption and regeneration. The book also considers the numerical approach for heat and mass transfer and applications, in addition to the challenges of nanofluids in industrial processes.

The book will be a useful reference for researchers and graduate students studying nanotechnology and nanofluids advancements within the fields of mechanical and chemical engineering.

Prof. Shriram S. Sonawane is a Professor of Chemical Engineering at Visvesvaraya National Institute of Technology [VNIT], Nagpur. M.S., India Dr. Sonawane's research area includes (a) polymer nanocomposites, (b) nanofluids, (c) nanoseparation technology, and (d) process modeling and simulation etc. He has more than nine patents to his credit. Dr. Sonawane is listed and recognized amongst the world's top 2% of scientists listed by AD scientific rankings Stanford University, USA. He has published more than 175 research papers in various national and international journals of repute. He was also nominated for the prestigious "Shanti Swaroop Bhatnagar Award". Dr. Sonawane has edited many books and published more than 75 book chapters in various reputable publishers' books. He was a Guest Editor for Sustainable Energy Technologies and Assessment. He was also the Executive Guest Editor of the *Current Pharmaceutical Bio-Technology* Journal ELSEVIER. He was also serving on the editorial body of *Research Journal of Chemistry and Environment*. Dr. Sonawane also served in the editorial body of the *World Research Journal*. He has received more than 5000 citations with h-index more than 40 and has also presented more than 150 research papers at various national and international conferences. Dr Sonawane is a recipient of Dr. Babasaheb Ambedkar Best Engineers Award by BANAE India. Moreover, he is awarded with best Innovative Researchers Scientist award. Dr Sonawane is a part of NBA, NAAC, UPSC, AICTE, SERB, DST and many Govt. of India supported bodies as expert. Dr Sonawane is associated with IIChE, MAS and he is a recipient of "Fellow" of these bodies. He served as expert on various Govt of India academic institutions of repute as well as State Universities of India. Dr Sonawane invited in many countries and delivered more than 75 key note addresses and invited talks. He also awarded with best paper presentations in many National and international conferences.

Dr. Parag P. Thakur is an Assistant Professor at SVNIT, Surat. He previously served as an faculty at NIT, Warangal. He completed his PhD from VNIT, Nagpur (India). He has received Post-doc research invitation from Chonnam National University, South Korea. His research areas are nanofluids, micro-fluids, nanocomposites, and modeling and simulation. He has published 15 research articles in reputed journals and received more than 400 citations for his research. He has published 36 book chapters and three patents, attended more than 50 international conferences all around India, and received the best presentation award at various international conferences. He is a reviewer of various journals like the *Journal of Mass and Heat Transfer* and *Separation Science and Technology*.

Emerging Materials and Technologies

Series Editor: Boris I. Kharissov

The *Emerging Materials and Technologies* series is devoted to highlighting publications centered on emerging advanced materials and novel technologies. Attention is paid to those newly discovered or applied materials with potential to solve pressing societal problems and improve quality of life, corresponding to environmental protection, medicine, communications, energy, transportation, advanced manufacturing, and related areas.

The series takes into account that, under present strong demands for energy, material, and cost savings, as well as heavy contamination problems and worldwide pandemic conditions, the area of emerging materials and related scalable technologies is a highly interdisciplinary field, with the need for researchers, professionals, and academics across the spectrum of engineering and technological disciplines. The main objective of this book series is to attract more attention to these materials and technologies and invite conversation among the international R&D community.

Non-Metallic Technical Textiles
Materials and Technologies
Mukesh Kumar Sinha and Ritu Pandey

Smart Micro- and Nanomaterials for Drug Delivery: Two-Volume Set
Edited by Ajit Behera, Arpan Kumar Nayak, Ranjan K. Mohapatra, and Ali Ahmed Rabaan

Smart Micro- and Nanomaterials for Drug Delivery: Volume One
Edited by Ajit Behera, Arpan Kumar Nayak, Ranjan K. Mohapatra, and Ali Ahmed Rabaan

Smart Micro- and Nanomaterials for Pharmaceutical Applications
Edited by Ajit Behera, Arpan Kumar Nayak, Ranjan K. Mohapatra, and Ali Ahmed Rabaan

Friction Stir-Spot Welding
Metallurgical, Mechanical and Tribological Properties
Edited by Jeyaprakash Natarajan and K. Anton Savio Lewise

Phase Change Materials for Thermal Energy Management and Storage
Fundamentals and Applications
Edited by Hafiz Muhammad Ali

Nanofluids
Fundamentals, Applications, and Challenges
Shriram S. Sonawane and Parag P. Thakur

For more information about this series, please visit: www.routledge.com/Emerging-Materials-and-Technologies/book-series/CRCEMT

Nanofluids
Fundamentals, Applications, and Challenges

Shriram S. Sonawane
Parag P. Thakur

CRC Press
Taylor & Francis Group
Boca Raton London New York

CRC Press is an imprint of the
Taylor & Francis Group, an **informa** business

First edition published 2024
by CRC Press
2385 NW Executive Center Drive, Suite 320, Boca Raton FL 33431

and by CRC Press
4 Park Square, Milton Park, Abingdon, Oxon, OX14 4RN

CRC Press is an imprint of Taylor & Francis Group, LLC

ISBN: 978-1-032-51987-6 (hbk)
ISBN: 978-1-032-51988-3 (pbk)
ISBN: 978-1-003-40476-7 (ebk)

DOI: 10.1201/9781003404767

Typeset in Times
by MPS Limited, Dehradun

Contents

Preface

Nanofluids: Fundamentals, Applications, and Challenges is an important book to understand the mathematical and experimental advances in nanofluid research. The main focus of the book is on the application of nanofluids for various sectors. The application of nanofluids in solar panels, car radiators, boiling, and CO_2 absorption is discussed in detail. This book is divided into three main sections:

1. Fundamentals of nanofluids (Chapter 1)
2. Application of nanofluids for the heat and mass transfer processes (Chapters 2–5)
3. Numeric approach of nanofluids application and challenges (Chapters 6–8)

The first section gives a detailed overview of the thermo-physical properties and efficiency enhancement of nanofluids in various operations. The second section gives a detailed overview of nanofluid application in solar panels, car radiators, boiling operations, and CO_2 absorption/regeneration and the third section provides an overview of the numeric approach for the mass transfer and heat transfer application of the nanofluids and challenges for the nanofluid applications in various sectors of industry. This book also provides recent advances and challenges of the nanofluid applications in industrial processes. This book is useful for researchers and professionals who are working in industry/academia or anyone interested in the applications of nanofluids in industrial processes for design purposes. The following sectors of the audience are the primary audience of the book

1. UG/PG/Doctoral students in the field of nanotechnology/chemical engineering and aspiring researchers within a broad domain of nanotechnology/chemical engineering.
2. Faculty and staff from reputed academic institutions and technical institutions.
3. Executive/engineer or researcher from manufacturing, service industry of chemical and biomedical industry.
4. Government organizations including R&D laboratories having dedicated research departments on nanotechnology and chemical processes.

Foreword

In recent years, research around nanofluids has increased due to their high thermal properties. The synergetic effect of two or more nanoparticles enables a wide range of applications in various industries. Studying the fundamentals and advanced topics related to nanofluids is important for students, researchers, scientists and faculties. *Nanofluids: Fundamentals, Applications, and Challenges* by Prof. Shriram Sonawane and Dr. Parag Thakur provides sufficient required knowledge about the topic of nanofluids and it is my pleasure to write the foreword for this book.

Before writing this foreword, I read the book in detail and I found the content of the book very useful for the students, faculties, and researchers working in this field. The first chapter is dedicated to the fundamentals of nanofluids. The synthesis of nanofluids, characterization techniques, and parameters affecting the stability of nanofluids are explained with recent data and studies. Chapters 2–5 are dedicated to the experimental study of car radiators, solar panels, boiling, and CO_2 absorption application of nanofluids, respectively. To discuss the application of nanofluids in heat exchange operation (heat transfer without phase change), solar panels and car radiators are used as a case study and whole chapters, i.e., Chapters 2 and 3 are dedicated to studying the parameters affecting the performance and efficiency of nanofluids. Heat transfer with phase change is disused using boiling operation as an example is discussed in Chapter 4 in detail. The concept is discussed in the case study of fly ash-based hybrid nanofluids. The use of nanofluids in mass transfer operation is discussed with the help of the CO_2 absorption example. Then Chapters 6 and 7 are dedicated to the numeric studies of the heat and mass transfer application of nanofluids, respectively. The last chapter, i.e., Chapter 8 is dedicated to the various applications of nanofluids and the challenges to nanofluids on a large scale. Examples of recent advances in hybrid nanofluids are presented in the book wherever necessary.

I find this book very interesting and useful for the students and researchers working not only in the field of chemical engineering but also for advanced heat transfer, nanotechnology, mechanical engineering, and numeric simulations. Thus, I believe this is a very important book for CRC Press of Taylor and Francis publications and it would be very beneficial for the students, researchers, scientists and faculties for conceptual understanding of this area.

Prof. Dr. Mohsen Sharifpur, PhD, Pr Eng.
Professor and Head for Nanofluids Research Laboratory
Department of Mechanical and Aeronautical Engineering
Engineering III, 6-85
University of Pretoria
Lynnwood Road
Pretoria, South Africa
Tel: +27 (0) 12 420 2448
Fax: +27 (0) 12 420 6632
E-mail: mohsen.sharifpur@up.ac.za

1 Introduction of Nanofluids

1.1 INTRODUCTION

Nanofluids are novel fluids consisting of suspension of nanoparticles of size 1–100 nm in base fluids like water, and amine. Generally, nanoparticles of metal oxides are used in nanofluids for better thermal performance. Water and other coolants are used as a base fluid for the heat transfer operation while amines are used for the CO_2 absorption operation [1–3].

The addition of nanoparticles leads to an increase in the thermal and physical properties of the base fluid. Thermal conductivity is an important thermal property in this regard. However, a decrease in the specific heat is generally observed after the addition of nanoparticles in the base fluid. Metal oxide-based nanofluids show comparatively better heat transfer performance. Different nanoparticles and base fluid and their respective thermal conductivity are shown in Figure 1.1. Thermal conductivity of MWCNT is 3000 W/mK. This value is comparatively very high than other metals and metal oxides. This increase in the thermal property is useful for various industrial heat transfer applications. However, an increase in the viscosity is also an important aspect of the use of nanofluids. Thus, the study of thermo-physical properties along with the rheological behavior of the nanofluids is an important aspect of the nanofluids.

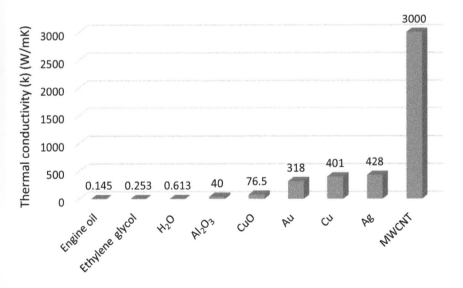

FIGURE 1.1 Different nanoparticles and base fluids and their thermal conductivity [4].

DOI: 10.1201/9781003404767-1

1

1.1.1 Important Features of Nanofluids

Nanofluids offer various advantages over the conventional coolants and solvents used for heat and mass transfer.

- Because of the smaller size of the nanoparticles, nanofluids offer a better surface area-to-volume ratio for heat and mass transfer. The smaller the diameter of nanoparticles better the performance of the nanofluids will be. Various researchers have shown the increased thermal conductivity and viscosity of nanofluids. For example, graphene nano platelets offer more thermal conductivity and viscosity compared to de-ionized water [4–6].
- Though nanofluids have a higher viscosity, nanofluids need less pumping power to perform equivalent heat transfer compared to conventional coolant. But pumping power increases with an increase in nanoparticle loading [7–9].
- The amount of nanoparticles used for the application is an important parameter for the nanofluids application. The physical properties of particular nanofluids can be modified by changing the amount of nanoparticles. These changed properties can be used for the suitable application of the nanofluids. These changed properties can be manipulated based on the calculated amount of nanoparticles. Surface wettability is an important factor for the application in the pool boiling processes. Which is primarily used for cooling reactors.
- Brownian motion is a very important aspect of nanofluids for the application as a heat transfer and mass transfer media Brownian motion of the nanoparticle contributes to the stability of nanofluids and thus the nanoparticles do not get agglomerated to reduce the thermal performance. This prevents the clogging of the pipes and transportation channels [10–12].

1.1.2 Disadvantages of Nanofluids

Nanofluids have a few disadvantages that need to be tackled. These disadvantages prevent the use of nanofluids on the industrial level.

- Synthesis of nanofluids is costly. Thus, it is difficult to synthesize large-scale nanofluids. The use of nanofluids on a large scale. The synthesis of nanoparticles like carbon nanotubes and MXenes is a costly process. The one-step approach of nanofluid synthesis is also a very costly affair.
- A higher amount of nanoparticles leads to the agglomeration of nanoparticles in the system. This can lead to clogging of the channel. However, the stability of nanofluids can be manipulated by various methods. However, it is very difficult to maintain the stability of nanofluids for a longer period.
- The use of nanoparticles in the base fluids increases the thermal performance of the system. But, in turbulent flow, the results are not promising in some cases.
- Nanofluids have more viscosity than conventional fluids. These require higher pumping power to operate. Thus, the use of nanoparticles can lead to the replacement of the existing pumping systems.

- Nanoparticle concentration is crucial for the nanofluids application. Thus, extra caution must be taken during the use of amount of nanoparticles. Slight changes in the nanoparticle amount may lead to changes in the thermo-physical properties and thus the efficiency of the nanofluids may affected.

1.2 PREPARATION METHODS OF NANOFLUIDS

Nanofluids can be synthesized in two ways. One method is a one-step method while the other method is called a two-step method. The amount of nanoparticles in the base fluid is an important parameter for the synthesis of nanofluids for the particular application. Various methods are represented in Figure 1.2. The use of surfactant can be done to improve the affinity between the base fluid and nanoparticles for the respective nanofluids application.

The following formula is generally used for the calculation of the amount of nanoparticles for the required concentration.

$$\emptyset = \frac{\left(\frac{W_{np}}{\rho_{np}}\right)}{\left(\frac{W_{np}}{\rho_{np}}\right) + \left(\frac{W_{bf}}{\rho_{bf}}\right)} \times 100 \qquad (1.1)$$

Here, the percent amount of nanoparticles is represented as $\emptyset$ and W_{np} is the nanoparticle's weight, ρ_{np} is the density of nanoparticles, W_{bf} is the base fluid weight, and ρ_{bf} is the base fluid density.

1.2.1 ONE-STEP METHOD

In the One-step method, nanoparticles and nanofluids are synthesized in a single step only. In this method, the simultaneous synthesis of nanoparticles within the

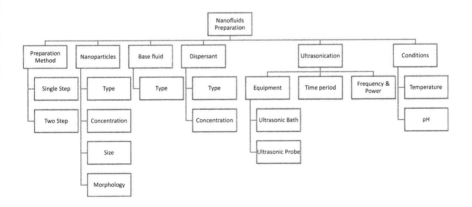

FIGURE 1.2 Different nanofluids synthesis methods [4].

base fluid is done. For example, In the Physical Vapor Deposition method, the direct evaporation and condensation of nanoparticles in the base fluid leads to the synthesis of the nanofluids. Compared to the two-step method, in this method, various steps can be eliminated. In the one-step method, synthesized nanoparticles have more uniform morphology and this leads to better stability than the two-step method [13]. Other methods like the Micro electrical discharge method (micro-EDM) are also used for the synthesis of Cu/ethylene glycol nanofluids. The synthesized nanoparticles had a 10 nm size with very little agglomeration [14,15].

- This method is more costly than the two-step method. Thus, it is difficult to synthesize nanoparticles using this method.
- In these methods, the nanoparticles may carry traces of harmful reactants after synthesis. Cleaning of the nanoparticles is not possible in this step [16–18].

1.2.2 Two-Step Method

In the two-step method, nanoparticles are synthesized separately. Then added to the base fluid. This method gives the flexibility to choose the synthesis method of the nanoparticles. The dispersion of the nanoparticles and base fluid is done by use of surfactant, ultrasonication, or direct mixing. Figure 1.3 represents the two-step synthesis method. Nanoparticles have surface activity in the fluid media. Thus, they tend to cluster. These clusters settle down at the bottom. Thus, the surfactants are used to prevent the formation of clusters. Surfactants are very important to prevent the sedimentation of the nanoparticles in the base fluid. Surfactants need to be added in suitable quantities. Excess surfactant leads to a decrease in the thermal property of the nanofluids while less amount of surfactant will not prevent the aggregation in the nanofluids. The important aspect of the nanofluids synthesis is the pH of the nanofluids. It is very important to maintain the pH of nanofluids in the

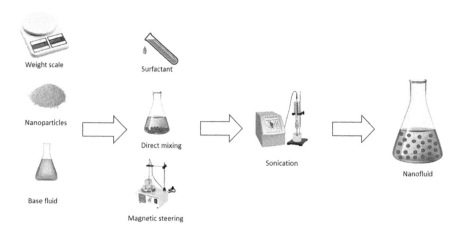

FIGURE 1.3 Step-wise synthesis route of two-step method [1].

system. The two-step method is widely used compared to the one-step method because of its simplicity. Hybrid nanofluids like alumina-copper/water are synthesized by this method. Firstly, water-soluble copper and aluminium nitrate are used to prepare the nanoparticles. During the synthesis of nanoparticles, spray drying of the nanoparticles is done. Then subsequent oxidation and reduction are done and finally, homogenization using hydrogen is done. Sodium lauryl sulfate is used as a surfactant in the base fluids to disperse nanoparticles then high-intensity ultra-sonication to achieve stable nanofluids [19–21]. But, in the two-step method stability generally reduces after 2 days of the synthesis.

1.3   THERMO-PHYSICAL PROPERTIES OF NANOFLUIDS

In various industrial operations, various heat transfer fluids (HTF) are used. These heat transfer fluids carry the heat energy for the exchange operation. Water or steam is used as a heat transfer fluid more often. Because, it is easy to available, cheap and does not react with most of the materials of construction. Along with water, sometimes ethylene glycol, oils are used in the water as an additive. This depends on the application of nanofluids. Because water has a comparatively lower boiling point. Thus, by using additives the boiling point of the working fluid is increased. The performance of nanofluids is measured by the relative increase in the thermal conductivity at the minimum increase in density and viscosity. This change in the thermo-physical property is important for the calculation of other heat and mass transfer properties. Thus, the nanofluids can effectively be used for waste energy recovery and process intensification of heat and mass transfer operations [22–26]. Heat transfer efficiency is solely dependent on the thermo-physical parameters of the heat transfer fluid. To manipulate the heat transfer efficiency, various strategies can be implemented. Changing the mechanical parts and processes are examples of an active approach while changing the fluid properties and changing the nanoparticle shape and size are passive approaches of improvising the heat transfer efficiency. Passive approaches are more preferred than active approaches because they offer more benefits and are less expensive. The use of nanofluids is an example of a passive approach to heat transfer enhancement [27,28].

1.3.1   THERMAL CONDUCTIVITY

The thermal conductivity of the nanofluids is a very important parameter in the thermo-physical parameters of the nanofluids because the amount of heat transfer is directly proportional to the thermal conductivity of the nanofluids in the system [29–34]. Thermal conductivity is directly proportional to the nanoparticle loading in the nanofluids. From the literature, it is evident that the increase in the Brownian motion is an important reason behind the increase in thermal conductivity [35,36]. Thermal conductivity is also dependent on the size and shape of the nanoparticles used in the nanofluids. The more the surface area of nanoparticle, the more thermal conductivity can be achieved. Nanoparticles with less particle size will have better thermal conductivity than nanoparticles with comparatively larger particle sizes for the same nanoparticle concentration. The shape of the nanoparticle is also important for heat transfer because the morphology with a higher surface area to volume ratio is better for the improved thermal efficiency of the same nanoparticle [37,38].

The temperature of the nanofluids is also a very important parameter for the change in values of thermal conductivity. Generally, the thermal conductivity of the fluids increases with the increase in temperature. Kedzierski et al. [39] reported the thermal conductivity of 0.2 W/m.K for the 0.1–0.4 vol% at 20–40 nm at 15–45 °C. Ohunakin et al. [40] reported an increase in the thermal conductivity by 2.75% and 0.45% in TiO_2-MO nanofluids for the nanoparticles of 13 nm size at 0.2 g/L concentration at 29–32 °C. Zawawi et al. [41] reported an increase of 2.41% in the thermal conductivity at 0.1 vol% concentration at 30 °C for 30 nm particle size.

Sanukrishna and Prakash [42] reported an increase in thermal conductivity of 1.48 W/m.K at a concentration of 0.6% at 20 °C at a concentration of 0.6 vol% with 13 nm size of nanoparticles. The authors also reported that the thermal conductivity of SiO_2-PAG nano-lubricant can be increased by 31% at 0.8 vol% compared to the pure lubricant at 21 nm size.

Zawawi et al. [43] showed that, at 30 °C, the thermal conductivity of 0.1% Al_2O_3-TiO_2/PAG nano-lubricants increases by 2.41%. Gill et al. [44] reported that, with a change in the nanoparticle loading from 0.2 g/L to 0.6 g/L, TiO_2-MO nano-lubricants increase in the thermal conductivity changes between 14.37% and 41.25%. The increase in the thermal conductivity can be seen in Figure 1.4.

Narayanasarma and Kuzhiveli [45] calculated the thermal conductivity of 0.2% SiO_2-POE nano lubricants as 1.109 W/m.K at 0.01–0.2 vol% at 5–20 nm size. Alawi et al. [46] reported an increase in the thermal conductivity by 28.88% at a 4% volume fraction at 35 °C. Various attempts are made to predict the thermal conductivity of the nanofluids using mathematical strategies. Two models are most prominently used for this purpose.

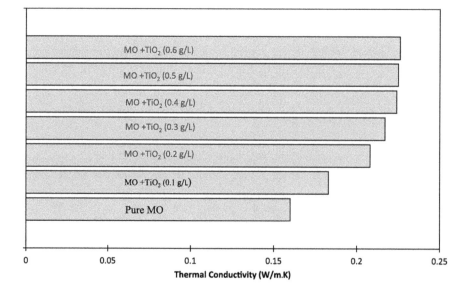

FIGURE 1.4   Effect of TiO_2 concentration on the thermal conductivity [44].

1. Wasp model
2. Maxwell model

For low nanoparticle loading, the Maxwell model serves the best results [47]. While Wasp model does not consider the morphology of the nanoparticles used thus, fails to predict the accurate value of thermal conductivity. Thus, the Maxwell model is most prominently used for thermal conductivity calculations [48].

1.3.2 Viscosity

The practical application of the nanofluids mainly depends on the viscosity of the nanofluids. Pressure drop and pumping power are directly dependent on the viscosity of nanofluids [49]. As the solids have more mass than liquids, the viscosity value increases for the nanofluids than the respective base fluid in all circumstances. The highly viscous fluid is not suitable for the system. Various researchers have reported an increase in the viscosity in various nanofluid systems [50]. Mahbubul et al. [51] used Al_2O_3-R141b nano-refrigerant for the study. The experiments were carried out with nanoparticle concentrations of 0.1–0.4% at a 5–20 °C temperature range. The highest viscosity value is obtained at 0.4 vol% concentration [45,46]. Some interesting findings were also reported by the other researchers. Nanofluids of TiO_2 with 10% nanoparticle concertation have reported a viscosity three times more than the base fluid and similarly, 10% nanoparticle concentration of Al_2O_3-based nanofluids had viscosities 200 times higher than the base fluids [52].

Kedzierski et al. [39] reported the viscosity decrease with the increase in the temperature. Authors have used the nanoparticles with 20–40 nm size and nanoparticle concentration of 0.1–0.4 vol% nanoparticle concentration at 15–45 °C. Ohunakin et al. [40] used silica and Titanium oxide as nanoparticles for the synthesis of nanofluids, in the case of SiO_2-based nanofluids, the viscosity increased by nearly 1%, while, in the case of TiO_2-based nanofluids, viscosity is increased by nearly 6%. The nanoparticle with 13 nm size and nanofluids of nanoparticle concentration 02 g/L are only used in both nanofluids synthesis. These studies were conducted at 29–32 °C.

Zawawi et al. [41] reported a viscosity increment of 20–50%, for the 0.02–0.1 vol% nanoparticle concentration with different nanoparticle sizes at 30–80 °C temperature.

Sanukrishna and Prakash [42] reported that nanofluids with 0.6 g/L nanoparticle concentration have a viscosity increase in the lubricant by 10 times than that of pure lubricants. Gill et al. [44] also reported the continuous increase in the viscosity with increasing nanoparticle concentration. Narayanasarma and Kuzhiveli [45] also reported that the increase in temperature leads to a decrease in the viscosity while increasing the nanoparticle concentration increases the viscosity. The increase in the scar during these experiments is shown in Figure 1.5. Wear scar radius (WSR) decreases in nanofluids than base fluid.

Alawi et al. [46] reported an increase in viscosity by nearly 13% for 4% nanoparticle concentration in nanofluids. Kumar et al. [53] reported an increase in the viscosity by 17% for the 1% nanoparticles at 60 °C. Various viscosity models developed so far are shown in Table 1.1.

(a) (b)

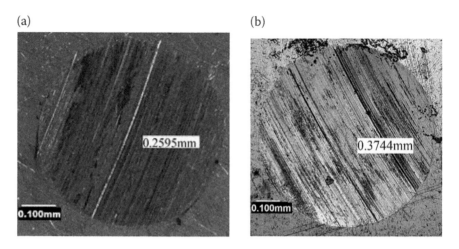

FIGURE 1.5 War scar radius for (a) nanofluids (b) base fluid [45].

TABLE 1.1
Viscosity Calculation Models for Nanofluids [45]

Model	Equation
Einstein	$\mu_{eff} = \mu_f (1 + 2.5\emptyset)$
Pak and Cho	$\frac{\mu_{eff}}{\mu_f} = 1 + 39.11\emptyset + 533.9\emptyset^2$
Brinkman	$\frac{\mu_{eff}}{\mu_f} = \frac{\mu_f}{(1-\emptyset)^{2.5}}$
De Bruijn	$\frac{\mu_{eff}}{\mu_f} = 1 + 2.5\emptyset + 4.698\emptyset^2$
Wang	$\frac{\mu_{eff}}{\mu_f} = 1 + 7.3\emptyset + 123\emptyset^2$
Maiga	$\frac{\mu_{eff}}{\mu_f} = 123\emptyset^2 + 7.3\emptyset + 1$
Gherasim	$\frac{\mu_{eff}}{\mu_f} = 0.904e^{14.8\emptyset}$
Mooney	$\frac{\mu_{eff}}{\mu_f} = 1 + 2.5\emptyset + [3.125 + (2.5/\emptyset_{max})]\emptyset^2$

1.3.3 DENSITY

Generally, an increase in the nanoparticle concentration leads to an increase in the density of nanofluids Alawi and Sidik [54,55] reported that an increase in the temperature also increases the nanofluids density. Authors have reported the results of density variation in the temperature range of 27–52 °C. Kedzierski et al. [39] also, studied the effect of temperature on nanofluids performance, and reported that the temperature increase leads to a decrease in the nanofluids density. Nanoparticle concentration is also changed from 1% to 5%. Mahbubul et al. [51] used a 5% nanoparticle concentration and reported an 11% enhancement in the density of

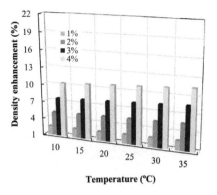
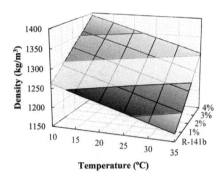

FIGURE 1.6 Increase in the density concerning increased temperature and concentration [46].

nanofluids. Alawi et al. [46] reported an 11.5% increase in the density at 4% nanoparticle concentration at 35 °C. The results of various base fluids and nanoparticles are shown in Figure 1.6.

1.3.4 Specific Heat Capacity

The energy required to increase the temperature of 1 kg of the sample by 1 K is called the specific heat capacity of the sample. Liquids require more energy to raise their temperature than solids. Increasing the nanoparticle concentration leads to a decrease in the specific heat of the nanofluids [47,48].

1.4 CHARACTERIZATION METHOD

Characterization of the nanoparticles and the nanofluids is required to know the exact values of different properties and parameters. As we have discussed in the previous section, particle size, the morphology of nanoparticles, the distribution of the nanoparticle and the stability of nanofluids are important parameters to develop an efficient nanofluids system. The morphology of nanoparticles, size, and stability of the nanofluids depends on the method of synthesis, pH of the base fluid, temperature of the base fluid, and surface treatment method used for the nanofluids synthesis [47,49,50,56].

1.4.1 Particle Size Measurement

Nanoparticle size can be determined by two methods: the first method is the microscopy method and the other method is the light scattering method. The microscopy method is useful when the shape of a nanoparticle is a matter of study. In the dynamic light scattering technique, the particle distribution in the nanofluids can be measured. The size measured in the electron microscopy may not be the overall size of the nanoparticle. While, the DLS technique, measures the hydrodynamic size

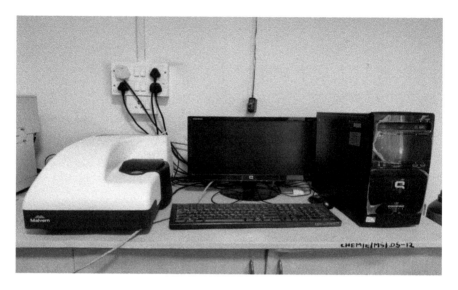

FIGURE 1.7   DLS equipment (Make: Malvern Panalyticals) [60].

of the nanoparticle in the nanofluids [57–60]. The zeta potential of the nanofluids is also determined by using DLS equipment. Figure 1.7 is a photograph of the dynamic light scattering equipment available in our laboratory at VNIT, Nagpur.

1.4.1.1   Dynamic Light Scattering Method

The DLS equipment traces the Brownian motion of the nanoparticles in the base fluid. Based on this movement the equipment measures the hydro-dynamic size of nanoparticles. In the nanofluids system, the surface of nanoparticles is in contact with various ions, adsorbed materials present in the base fluids or the surfactant used to improve the stability of nanofluids. Nanoparticles with higher weights move slowly compared to the nanoparticles with the lighter weight. This diffusion of the nanoparticles in the base fluid is called translational diffusion. DLS equipment measures the translational diffusion of the nanoparticles and calculates the hydrodynamic size of the nanoparticles present in the base fluid. The hydro-dynamic size is the size of spherical particles with the same translational diffusion. This calculation is based on the Stokes-Einstein equation. The equation is represented as follows in Equation 1.2:

$$D_H = \frac{kT}{3\pi\eta D} \tag{1.2}$$

Here, D_H is the nanoparticle hydrodynamic diameter; D is the translational diffusion coefficient; T is absolute temperature; η is the viscosity of the nanofluids; while Boltzmann's constant is represented by k; Figure 1.8 is the graph obtained from the DLS equipment for the size distribution in the sample [61–63]. The hydrodynamic size of the nanoparticles measured in the figure 1.8 is 90 nm.

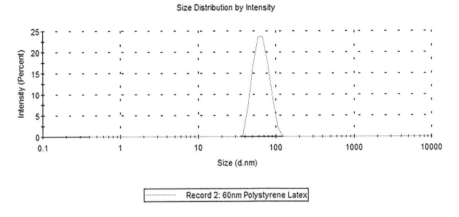

FIGURE 1.8 Size distribution of the polystyrene latex obtained from DLS equipment [60].

1.4.1.2 Microscopic Analysis

The following three methods are generally used for the size calculation of the nanofluids [64–66]:

1. Atomic Force Microscope (AFM)
2. Scanning Electron Microscope (SEM)
3. Transmission Electron Microscope (TEM)

Figure 1.9 represents the various shapes of TiO_2 nanofluids reported by Azmi et al. [67]. They observed the FESEM and TEM images of the TiO_2 nanoparticle and TEM images of the nanofluids.

1.4.2 THERMAL CONDUCTIVITY ANALYSIS

As we have discussed in Section 1.3, nanofluid's thermal conductivity is arguably the most important among all the other thermo-physical properties for the heat transfer increment application.

Researchers have pointed out that, various parameters contribute to the increase in the nanofluids performance for the heat transfer application. Ambreen et al. [68] studied the various nanoparticles with different sizes and their effect on thermal conductivity. Researchers have preferred the nanoparticle size should be less than 100 nm and various factors affecting the effective thermal conductivity of nanofluids are shown in Figure 1.10. Important parameters like base fluid properties, thermal conductivity analysis methods are highlighted in the Figure 1.10.

Ahmadi et al. [69] have reviewed the various studies to measure the effect of temperature on the thermal conductivity. The authors also pointed out that, the nanoparticle loading is directly proportional to the heat transfer enhancement. Morphological analysis is important for the viscosity and viscosity ratio [70]. However, there is scope for detailed research on the effect of morphological data on thermal conductivity. Because various researchers have conflicting data on this aspect.

(a) FESEM of TiO$_2$ nanoparticles at X 50,000
magnifications

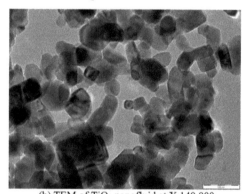

(b) TEM of TiO$_2$ nanofluid at X 140,000
magnifications and 1.5 % volume concentration

FIGURE 1.9   Morphological analysis of TiO$_2$ nanoparticle and TiO$_2$/water nanofluids [67].

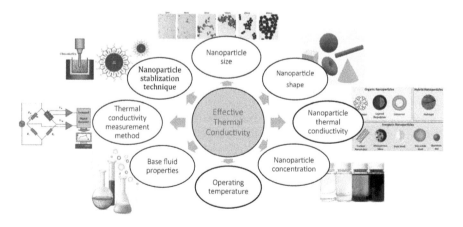

FIGURE 1.10   Various parameters affecting the effective thermal conductivity [68].

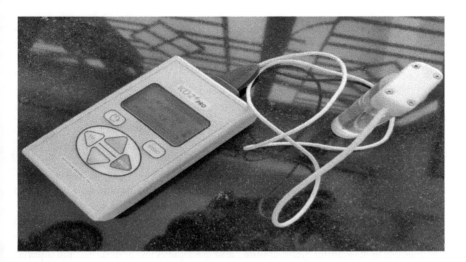

FIGURE 1.11 KD-2 pro thermal conductivity analyzer (Make: Decagon Inc, USA) for thermal conductivity measurement [60].

In recent years, the research with the hybrid nanofluids has increased compared to the mono-nanoparticle-based nanofluids. The synergetic effect of the nanoparticles leads to offering better heat transfer efficiency than other mono-nanoparticle-based nanofluids. Sajid et al. [71] did experimentations with the various hybrid nanofluids systems and reported that the hybrid nanofluids offer better heat transfer than the nanofluids with single nanoparticles.

ISO and ASTM have developed various standard methods to measure the thermal conductivity of nanofluids. For example, ISO has developed three major standard methods to measure thermal conductivity, i.e., ISO 8497, ISO 8301, and ISO 22007-2:2008. ASTM also developed several standard methods to measure thermal conductivity, e.g., ASTM Standard D2717-95 ASTM Standard D5470-06, and D5334-14. The transient hot wire method is used more frequently than other methods for thermal conductivity calculations. This is a steady-state type of method for the thermal conductivity analysis. A known amount of heat flux is applied at one end of the needle, this flux is generally generated by the electrical voltage. The surface area for the heat transfer and needle thickness are constant. After inserting the needle into the sample. We have to wait for the steady-state temperature condition. The difference in temperature along the length of the needle is measured by the thermo-couples mounted in the needle. This method is used in various thermal conductivity analyzers available in the market. Figure 1.11 shows the KD2 Pro thermal conductivity analyzer.

1.4.3 Specific Heat Analysis

Generally, the specific heat increases with the temperature rise. Thus, the calorimeters can be used to measure the specific heat of the nanofluid sample. Specific heat of the nanofluids is measured by the differential scanning calorimeter

(DSC) technique. Generally, the ASTM method of E 1269-05 is used for the specific heat analysis. In DSC, two pans are required for the sample analysis. The first pan is the empty pan for the baseline and the other pan contains the sapphire with known specific heat value. Then the nanofluids are put in the reference pan by using the sapphire-specific heat value for the particular temperature as a reference. The value of specific heat for the nanofluids is calculated. Calorimeters use the following equation to determine the specific heat of the sample.

$$C_p = C_{p,st} \cdot \Delta q_s \cdot m_{st} / \Delta q_{st} \cdot m_s \tag{1.3}$$

C_p is nanofluid's specific heat value; $C_{p,st}$ is the specific heat value of sapphire; Δq_s is the difference between a sample and an empty pan heat flow [72].

Specific heat capacity is important for the thermo-dynamic study of the energy and exergy calculations of the heat transfer operations. Specific heat is dependent on the nanoparticle loading, temperature, and nanoparticles and base fluid interaction [73].

Studies on the specific heat capacities of the hybrid nanofluids are reported by various researchers. Moldoveanu et al. [74] reported that the synergetic effect of the various nanoparticles has a large impact on the thermal behaviour of the hybrid nanofluids and a large scope of the research is present in this field. The author has compared the specific heat of the various single nanoparticle-based nanofluids and respective hybrid nanofluids. They found that the hybrid nanofluids have comparatively better specific heat capacity. This is due to the synergetic impact of the nanoparticles. As mentioned before, the mathematical models can be derived for the various nanoparticles and hybrid nanofluids systems. Figure 1.12 is a photograph of the differential scanning calorimeter equipment available in our laboratory at VNIT, Nagpur that we use for the specific heat measurement.

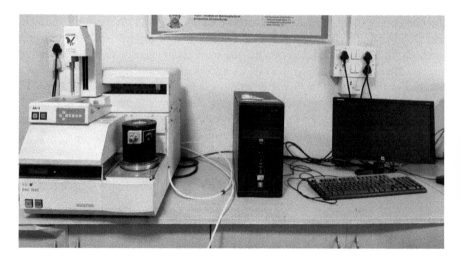

FIGURE 1.12 DSC machine available in VNIT, Nagpur for the specific heat measurement [60].

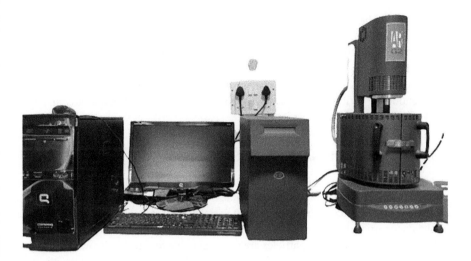

FIGURE 1.13   Real-time images of the rheometer available in our laboratory at VNIT, Nagpur for the viscosity measurement of nanofluids [60].

1.4.4   Viscosity Measurement

As we have discussed in the previous section, the viscosity always increases due to the addition of the nanoparticles. Various parameters are important for the viscosity of nanofluids like shape, size, and concentration of nanoparticles used.

In most cases, even with comparatively higher viscosity, nanofluids exhibit better thermal performance than base fluids [75]. An increase in the temperature leads to a decrease in the viscosity of nanofluids. Other parameters like particle size, shape of nanoparticles and pH of the nanofluids may have an impact on the viscosity of nanofluids. However, research on this has no definite conclusions.

Figure 1.13 is a real-time image of a rheometer available in our laboratory at VNIT, Nagpur for the viscosity measurement of nanofluids.

1.4.5   Density Analysis

Like viscosity, the density of the nanofluids also increases with the increase in the amount of nanoparticles. The following equation is generally used to calculate the density of nanofluids.

$$\rho_{nf} = \left(\frac{\varnothing}{100}\right)\rho_p + \left(1 - \frac{\varnothing}{100}\right)\rho_f \tag{1.4}$$

Here, ρ_{nf} is nanofluids density, ρ_p is nanoparticle density, ρ_f is the base fluid density.

$$\varnothing = \frac{m_p}{m_p + m_f} \times 100 \tag{1.5}$$

In the case of heat transfer by natural convection, the density of the material is a very important parameter. Generally, the density of the nanofluids can be measured by the density bottles available in the laboratories. Some interesting findings were reported by Chavan and Pise [76]. Authors have studied the density of Al_2O_3 /water and reported that these nanofluids behave as Newtonian fluids. Mariano et al. [77] also conducted various experiments to study the SnO_2-based nanofluids and they showed non-Newtonian behaviour.

1.5   STABILITY OF NANOFLUIDS

Nanoparticles in the nanofluids tend to form clusters of nanoparticles. This is due to the different charges present on the nanoparticle's surface. These various charges attract each other. These forces are present due to various oppositely charged groups present on the nanoparticle surface. These bigger size clusters change the physical properties of the nanofluids as the nanoparticles are no longer able to travel to Brownian motion. In the absence of the Brownian motion, these particles form sediment. Thus, it is very important to make nanofluids stable for better thermo-physical properties.

Generally, to increase the stability of nanofluids various strategies are implemented. Like the use of surfactant for surface treatment, the ultrasonication technique and manipulation of pH are used. Suitable surfactants or dispersants are used to stabilize the nanofluids. Ultrasonication is preferred over the surfactant, as the use of surfactant may decrease the thermal conductivity of nanofluids and ultrasonication doesn't change the thermo-physical properties.

1.5.1   ADDITION OF SURFACTANTS

The use of surfactants is a commonly used method to stabilize the nanofluids. Surfactants increase the wettability of the nanoparticles in the base fluids. Thus the stability of the nanofluids increases. Surfactant develops similar charges on the surface of nanoparticles, which leads to an increase in the zeta potential value of nanofluids. Surfactants are generally not very costly and they are easy to use. Surfactants have a hydrophilic head with a long hydrophobic chain. Although, surfactants can increase the stability of nanofluids. But, surfactants also can decrease the heat transfer performance of the nanofluids. Surfactants can produce the foam during the turbulence or heating of the nanofluids. Sodium dodecyl benzoic sulfate (SDBS) is the most commonly used surfactant for various nanofluids [77–79]. Researchers have studied the optimum concentration of the surfactant to be used for the nanofluids synthesis. Thus, generally, a surfactant of 25% of the nanoparticles concentration is used for the nanofluids synthesis. Various nanofluids and their particle size distribution graph are represented in Figure 1.14. This shows that the surfactants improve the stability of nanofluids.

1.5.2   FUNCTIONALIZATION OF NANOPARTICLES

Another method of improving the stability of nanofluids is functionalizing the nanoparticle surface with a suitable surface treatment method. The objective of the

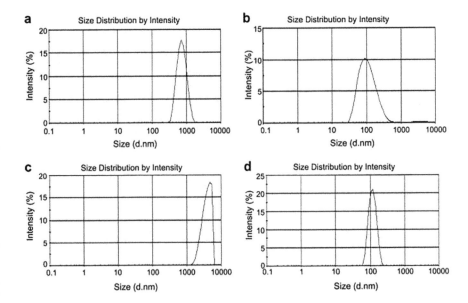

FIGURE 1.14 DLS results obtained for the (a) alumina/water nanofluids without surfactant, (b) alumina/water nanofluids with water, (c) copper/water without surfactant, (d) copper/water with surfactant [80].

surface treatment is to improve the nanoparticle and base fluid interaction, to reduce the attractive forces between two molecules, and to reduce the nanoparticle's attractive forces. The functionalization should be done properly otherwise the surface of the nanoparticle may get contaminated and the dispersion of nanofluids may be reduced [81].

1.5.3 USE OF ULTRASONICATION

Ultrasonication is one of the widely used methods for nanoparticle dispersion in nanofluids. The sonic waves induced from the sonicators develop the static charges on the nanoparticle surface. These charges are repulsive and thus, the zeta potential of nanofluids increases. Thus the stability of the nanofluids increases. The advantage of ultrasonication is that it does not compromise the thermal properties of the nanoparticles. However, many researchers use both techniques to synthesize the nanofluids. Sonication can be done directly using probe-type sonicators of different shapes and power or indirect sonicators like bath-type sonicators are used. Probe-type sonicators can develop the localized heating of the probe. This can lead to cavitation of the probe surface and thus the erosion of the probe surface may take place. This eroded surface will not deliver the required sonication for the nanofluids. Thus, it is very important to prevent the localized heating of the probe-type sonicators [82]. Figure 1.15 represents the effect of sonication time and morphology of synthesized nanoparticles.

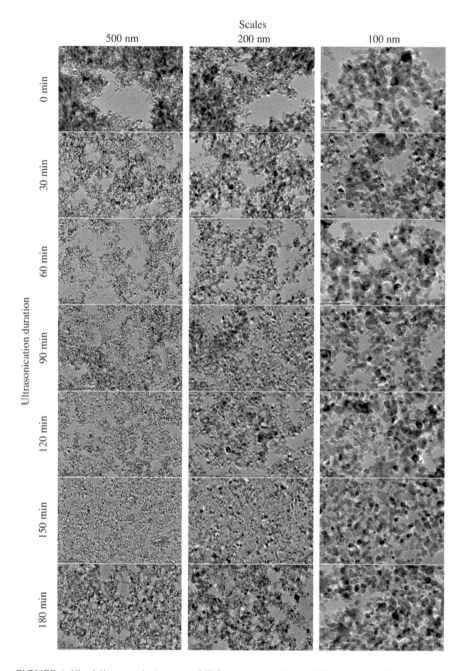

FIGURE 1.15   Microscopic images of TiO$_2$ nanoparticles at different times [82].

1.5.4   MANIPULATION OF pH VALUE

The stability of nanofluids is measured in terms of zeta potential at a particular pH value. Zeta potential without pH value has no meaning. For stable nanofluids, the

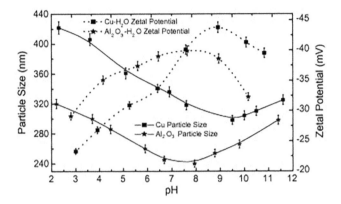

FIGURE 1.16 Effect of pH of nanofluids on nanoparticle size and zeta potential of nanofluids with SDBS [80].

zeta potential should be below -30 mV or above $+30$ mV. The solution with the zeta potential below -30 mV and above $+30$ mV is known as the highly stable solution. And nanofluids with values between -30 mV and $+30$ mV are considered less stable solutions. However, the zeta potential can be manipulated by the addition of acidic (positively charged) or basic (Negatively charged) components in the nanofluids. The pH value at which certain solutions have the zero zeta potential is known as the iso-electric point of the solution. If we keep adding the positive or negatively charged ions in the solution, the zeta potential will move away from the iso-electric point, on the positive side or negative side of the pH scale. Figure 1.16 represents the effect of pH on nanoparticle size and zeta potential value of nanofluids.

1.6 PARAMETERS INFLUENCING THE STABILITY OF THE NANOFLUIDS

The stability of nanofluids is dependent on the various parameters. In this section, a detailed discussion of these parameters is given.

1.6.1 THE DIELECTRIC CONSTANT OF BASE FLUID

The dielectric constant is the ability of the material to store the electric energy. The dielectric constant increases with the increase in the repulsive forces present on the nanoparticle surface. Thus, the larger the dielectric constant larger the repulsive forces between the nanoparticles and thus, the better the stability of nanofluids. Compared to other base fluids, water has a comparatively higher dielectric constant. Thus, water is most suitable for the stable nanofluids.

1.6.2 SURFACE CHARGE

Generally, the surface of the nanoparticle is surrounded by various charges. These charges can be divided into two regions. The layer attached to the nanoparticle surface is the stern layer. Charges in this layer are firmly attached to the nanoparticle surface. Another layer of charge is known as a diffused layer. Charges present in this layer are loosely attached to the nanoparticle surface. These two layers of the charges are collectively known as the Electrical double layer (EDL). The net charges present in the electrical double layer are termed as zeta potential. A value of the zeta potential should be more than +30 mV or below −30 mV. Otherwise, the nanofluids are considered less stable. Even with suitable zeta potential values, nanofluids may become unstable at higher densities [79,80].

1.6.3 pH Value

The pH value of the nanofluids is the most important parameter for the stability of nanofluids. Thus, during the nanofluids application, one must keep in mind to keep the pH of nanofluids stable. Zeta potential value is directly dependent on the pH of the nanofluids. Several researchers [83,84] reported the alteration of pH has an immediate effect on the stability of nanofluids. Charge density present on the nanoparticle surface is dependent on the pH value of nanofluids. Figure 1.17 represents the Iron oxide nanoparticles-based nanofluids and their zeta potential at respective pH values. At the iso-electric point, the nanoparticle surface does not carry any charge; i.e., the zeta potential value is zero [85]. Nanofluids having a zeta potential value near the iso-electric point are considered very unstable. Thus, while synthesizing the nanoparticle for the nanofluids system, it should be kept in mind that, the resulting nanoparticle may affect the pH of the nanofluids and ultimately the stability of nanofluids. Li et al. [86] used silver nanoparticles to investigate the effect of pH on the stability of nanofluids. At a pH of 11, the nanoparticles of uniform size are synthesized with the spherical particle shape. But, when the pH of nanofluids is decreased to 7, the shape of the nanoparticle gets changed. The shape and size are not of the uniform. In the following section, a detailed discussion of the effect of particle size and morphology on the stability of the nanofluids is given.

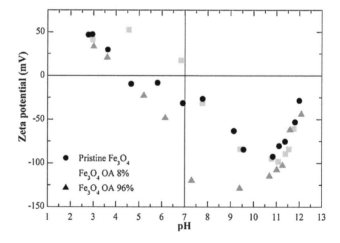

FIGURE 1.17 Variation of the zeta potential value at various pH values for different nanofluids [83].

1.6.4 MORPHOLOGY OF THE PARTICLES

For particles with nano size, more active sites are present on the particle surface. Thus, the reduction in the particle size also increases the active sites present on the nanoparticles. This increases the density of active sites in the nanofluid system [87]. He et al. [88] reported that, the ability to form agglomerate increases with the smaller size of nanoparticles. The author has used hematite ($-Fe_2O_3$) nanoparticles to conduct the study. Aggregation of smaller particles occurs more frequently even if the other parameters like pH and ionic strength are the same. As the nanoparticle size decreases, the coagulation loading and the iso-electric points fall to lower values. As a result of this, the agglomeration regime moves in the direction of a lower pH value [89].

Kim et al. [90] conducted an experimentations with the alumina/water-based nanofluids. This study showed that, with passing time, the stability of the nanofluids decreases and thus the shape of nanoparticles also changes. These different shapes of nanoparticles are shown in Figure 1.18. Among these different shapes, blade-

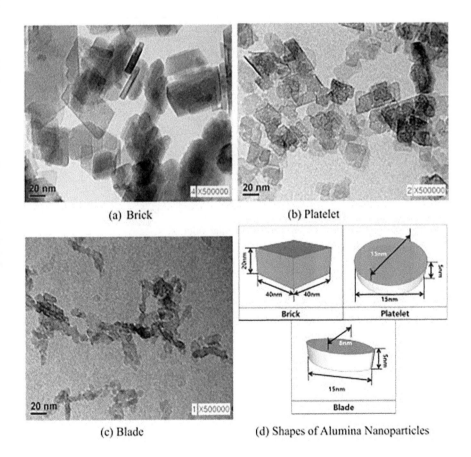

(a) Brick   (b) Platelet

(c) Blade   (d) Shapes of Alumina Nanoparticles

FIGURE 1.18 Morphological images of alumina nanoparticles using TEM [90].

shaped particles settle more quickly than Platelet and brick-shaped particles. Bricks-like shaped nanoparticles tend to be suspended in the nanofluids. The attractive forces between the nanoparticles are dependent on nanoparticle concentration and no such study conducted on the effect of nanoparticle morphology on the attractive forces. But, the shape of the nanoparticle does affect the repulsive forces of the nanoparticles. The brick-shaped nanoparticles have the highest repulsive potential and blade-like nanoparticles have the lowest nanoparticle repulsive force. These studies are shown in Figure 1.19. The blade-like nanoparticles get settled more quickly than the blade-like nanoparticles.

The formation of clusters is less likely in spherical nanoparticles than the isometric nanoparticles. If the most of particles are perpendicular in cylindrical nanoparticles, then the interaction potential is much less compared to the parallel arrangement of the cylindrical nanoparticles. Nanoparticles with higher aspect ratio have more tendency to form the agglomeration in the nanofluids. The relation of setting velocity and the operational parameters is represented in Equation 1.6;

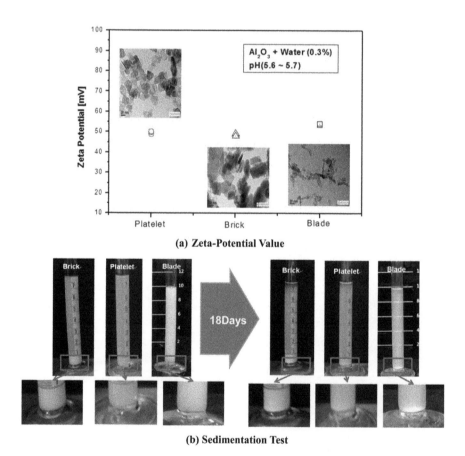

(a) Zeta-Potential Value

(b) Sedimentation Test

FIGURE 1.19 Sedimentation/photograph method results for the different nanoparticle shapes [90].

the particle shape directly influences the drag coefficient. Thus, the small nanoparticles with the minimum density difference are very useful to maintain the stability of the nanofluids.

1.6.5 PARTICLE CONCENTRATION

If the nanoparticles are added in higher concentration the average distance between the nanoparticles gets reduced significantly. This reduced distance between the nanoparticle surfaces leads to the development of the van der Waal force of attraction. When this van der Waal force of attraction balances the repulsive force of the nanoparticles, the stability of nanoparticles starts to decrease. Thus, it is very important to find the optimum particle concentration for the respective nanofluids system. Thus, the nanofluids with very high nanoparticle concentration have less stability than the nanofluids with low nanoparticle concentration. However, the higher nanoparticle concentration offers better heat transfer and mass transfer efficiency [91,92]. Various researchers have shown that the agglomeration increases with the increase in nanoparticle concentration. Chakraborty et al. [93] have reported an increase in the cluster size from 86 nm to 126 nm. The findings of this study are represented in Figure 1.20. Nanoparticles have shown highest cluster size of 125 nm for 0.8 vol % concentration. An increase in the nanoparticle cluster also depends on the time passed. Hong et al. [94] reported that, for a 0.2 vol% concentration of iron/ethylene glycol nanofluids system, the average cluster size of Fe/EG nanofluids increased from 1.2 micrometres to 2.3 micrommetres with time. This increase is quicker for the higher-concentration nanoparticles than lower-concentration nanoparticles.

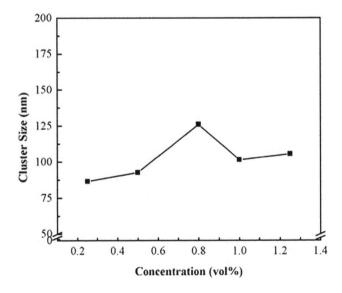

FIGURE 1.20 Nanoparticle loading vs nanoparticle cluster size [92].

1.7   STABILITY MEASUREMENT METHODS FOR NANOFLUIDS

The measurement of nanofluid's stability is crucial for the understanding of the nature of nanofluids for practical application. The accurate measurement of the stability can prevent the wastage of resources like time and money. There are few nanofluid stability methods available in the literature. The zeta potential measurement method is the most widely known and used method of all. Apart from that, less accurate, simple methods like the sedimentation method and absorbance method are used for the stability evaluation of nanofluids. The 3-omega method is also less known and very rarely used method for the stability measurement of nanofluids.

1.7.1   SEDIMENTATION/PHOTOGRAPH METHOD

In this method, the nanofluids are stored in the container in an unhindered place. The photographs of these containers are taken at regular intervals of time. These containers should not be hindered at any point in time. The nanoparticles present in this container are allowed to settle at the natural settling velocity. From the photographs of the nanofluids, we have to evaluate the comparatively more stable nanofluids. This method is easiest but very time-consuming and inaccurate. This method does not give any concrete data regarding the stability of nanofluids. But, still, various researchers have used this method to study the stability of nanofluids [96]. As shown in Figure 1.19 the stability evaluation can be done by using the sedimentation technique.

Stokes's law is used to calculate the settling velocity (V), this law is shown in the equation 1.6 [80],

$$V = \frac{2R^2}{9\mu}(\rho_p - \rho_L) \cdot g \tag{1.6}$$

Here, the radius of the particle is represented as R, the viscosity of the liquid medium is shown as μ and the density of the particle and liquid is represented by ρ_p and ρ_L, respectively. Thus, generally, to decrease the nanoparticle settling velocity, we need to decrease the nanoparticle size. But to increase in the viscosity is more challenging than the use of smaller size nanoparticles. Thus, smaller-size nanoparticles are always preferred for the enhancement of nanofluid stability [97]. Figure 1.21(a) is a schematic representation of the sedimentation/photograph technique of nanofluid stability evaluation.

The time consumed during the sedimentation of the nanoparticles by gravitational force can be reduced by using the centrifugation method. We need to centrifuge the nanofluids at the same speed and for the same time. Then the relative difference between the settled nanoparticles can be observed after a fixed interval of time. Various researchers use this technique to observe the nanofluid's stability. The centrifugation method drastically reduces the time consumed by the sedimentation method. Singh et al. [98] used the centrifugation method for the evaluation of the stability of silver/ethanol in the presence of poly-vinyl-pyrrolidone as a surfactant and reported 10 hours of stability even under 3000 rpm centrifugation.

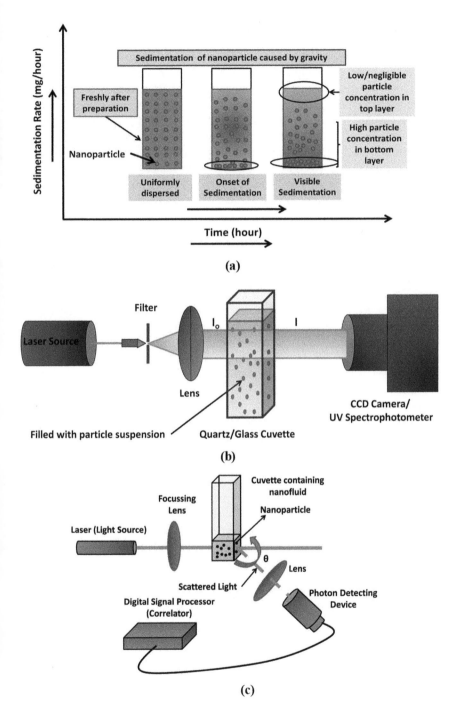

FIGURE 1.21 (a) Sedimentation/photograph method, (b) spectral absorption method, (c) dynamic light scattering method [95].

1.7.2 SPECTRAL ABSORBENCY ANALYSIS

The spectral absorbance of the nanofluids is measured at regular intervals of time. A spectrophotometer of the ultraviolet light is used for this purpose. Beer-Lambert law is used for the calculation of the nanoparticle concentration. If the nanoparticle concentration in the nanofluids decreases with time, then nanofluids are considered unstable nanofluids [99]. The disadvantage of this method is that it cannot be used for all the nanofluid systems. We must know the exact absorbance value of the nanoparticles used in the nanofluids. This method detects the samples having absorption peaks of 190–1100 nm [100]. These samples are tested at a regular interval, a decrease in the absorbance peaks represents a decrease in the nanofluids stability [82]. The spectral absorbance method is shown in Figure 1.21(b).

1.7.3 SURFACE CHARGE MEASUREMENT METHOD

Zeta potential analysis determines the total charges present on the nanoparticle surface. The more the different charges present on the surface, the more repulsive force will be there. Thus, the nanoparticles will not form a cluster of agglomeration. Thus, the zeta potential is a good parameter to measure the stability of the nanofluids. Fedele et al. [101] conducted a comparative study of the zeta potential values for the various mono-nanoparticle-based nanofluids. Authors have used nanofluids made up of various nanoparticles like copper oxide, Titanium Oxide and single-walled carbon nano-horns. These nanoparticles are also synthesized by various synthesis methods. Researchers reported that the synthesis method is also an important factor for the zeta potential of the nanofluids. These nanofluids are synthesized by three methods; in the first method of synthesis, nanoparticles are subjected to sonication at 130 W and 20 kHz for 1 hour. In the second method, nanoparticles are homogenized at a high pressure of 1000 bar and in the third method, a ball milling performed at the speed at 300 rpm for 2 hours [101]. Zeta potential values are measured by the dynamic light scattering method. DLS results for the zeta potential are represented in Figure 1.22 Zeta potential value is -80 mV is obtained from the graph.

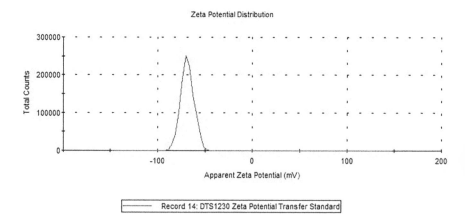

FIGURE 1.22   DLS results for the zeta potential of nanofluids [59].

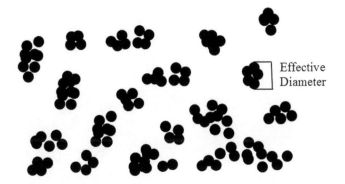

Effective Diameter

FIGURE 1.23 Clusters of nanoparticles present in the nanofluids [82].

The charge developed on the interphase of the liquid medium and nanoparticle surface is called a zeta potential. At the interphase, the total charge is distributed in two layers. The charge attached to the stern layer attracts the opposed charged particles. These charges are loosely attached to the stern layer. These charges form the second layer. This layer is called a diffused layer of charges.

The zeta potential is measured by the dynamic light scattering equipment. The refractive index and absorbance value of the sample should be known for the zeta potential analysis. With these values, the predefined program calculates the zeta potential values based on translational diffusion velocity. The cuvettes used for the zeta potential apply the potential difference across the cuvette. These cuvettes are U-type capillary tubes. Nanoparticles move towards these applied potential differences. Particles with high zeta potential will move more quickly toward the respective potential point. The machine calculates the translational diffusion of the nanoparticles and, thus, determines the zeta potential value. The detailed discussion on zeta potential and its significance is already discussed in the previous sections. By using various strategies we can manipulate the zeta potential of the nanofluids.

The zeta potential is heavily influenced by the pH of the nanofluids. The value of zeta potential can be easily manipulated by the pH alteration. pH of the nanofluids is maintained at the value away from the iso-electric point. The addition of the salts decreases the zeta potential of the nanofluids, as salts overshade the charges present in a double layer of charges present on nanoparticles. The schematic representation of zeta potential analysis using the dynamic light scattering technique is represented in Figure 1.21(c). But, the disadvantage of this method is that the even cluster size can be measured as the effective size of nanoparticles. Figure 1.23 is the representation of such clusters formed in the nanofluids.

1.7.4 3-OMEGA METHOD

As we have discussed in the section on thermophysical properties, changes in the thermos-physical properties are important characteristics of the nanofluids. If nanofluids are stable if nanofluids are stable, then, these properties differ over the

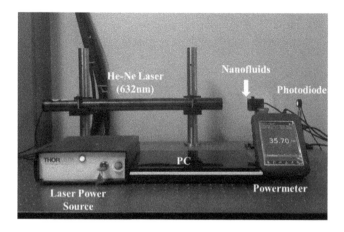

FIGURE 1.24   Laser scattering equipment of nanofluids stability evaluation [90].

period. In the 3-omega method, the thermal conductivity of nanofluids is measured at regular intervals of time. If the nanofluids are stable then the thermal conductivity of the nanofluids will not change. But, for the unstable nanofluids, the thermal conductivity will change significantly. Thermal conductivity is relatively easier to measure than the specific heat of the nanofluids. Thus, the thermal conductivity analysis is done for the stability of nanofluids. Other methods like laser scattering equipment are also used for the nanofluid stability evaluation. The la-scale setup of laser scattering equipment is shown in the Figure 1.24.

1.8   CONCLUSION

In this chapter, the nanofluids and their synthesis methods are described in detail. Nanofluids have various advantages over the conventional coolants used for heat transfer and solvents used for the mass transfer operation. The main challenge for the use of nanofluids on a large scale is the stability of nanofluids. Various factors affecting the stability of nanofluids are discussed in detail. Various methods for the evaluation of stability and to increase the stability are discussed in detail in this chapter. The application of the nanofluids in the heat and mass transfer operation is discussed in detail in the next chapters.

REFERENCES

1. A. R. I. Ali, & B. Salam (2020). A review on nanofluid: Preparation, stability, thermophysical properties, heat transfer characteristics and application. *SN Applied Sciences*, *2*(10), 1–17.
2. M. Malika, & S. S. Sonawane (2022). The sono-photocatalytic performance of a Fe_2O_3 coated TiO_2 based hybrid nanofluid under visible light via RSM. *Colloids and Surfaces A: Physicochemical and Engineering Aspects*, *641*, 128545.
3. S. B. Potdar, P. Saudagar, I. Potoroko, U. Bagale, S. Sonawane, & S. H. Sonawane (2022). Recent advances and reports on encapsulation in the food matrix: A review. *Current Pharmaceutical Biotechnology*.

4. N. Sezer, M. A. Atieh, & M. Koç (2019). A comprehensive review on synthesis, stability, thermophysical properties, and characterization of nanofluids. *Powder Technology, 344*, 404–431.
5. R. S. Khedkar, S. S. Sonawane, & K. L. Wasewar (2013). Synthesis of TiO$_2$-water nanofluids for its viscosity and dispersion stability study, *Journal of Nano Research, 24*, 26–33.
6. M. Mehrali, E. Sadeghinezhad, M. A. Rosen, S. T. Latibari, M. Mehrali, H. S. C. Metselaar, & S. N. Kazi (2015). Effect of specific surface area on convective heat transfer of graphenenanoplatelet aqueous nanofluids. *Experimental Thermal and Fluid Science, 68*, 100–108.
7. J. Yang, X. Yang, J. Wang, H. H. Chin, & B. Sundén (2022). Review on thermal performance of nanofluids with and without magnetic fields in heat exchange devices. *Frontiers in Energy Research, 10*, 1–25.
8. S. Noor, M. M. Ehsan, S. Salehin, & A. S. Islam (2014). Heat transfer and pumping power using nanofluid in a corrugated tube. *Heat Transfer, 8*, 11.
9. S. S. Sonawane, P. P. Thakur, M. Malika, & H. M. Ali (2022). Recent advances in the applications of green synthesized nanoparticle based nanofluids for the environmental remediation. *Current Pharmaceutical Biotechnology, 24*(1), 188–198.
10. P. Thakur, S. Sonawane, I. Potoroko, & S. H. Sonawane (2021). Recent advances in ultrasound-assisted synthesis of nano-emulsions and their industrial applications. *Current Pharmaceutical Biotechnology, 22*(13), 1748–1758.
11. R. K. Shukla, & V. K. Dhir (2008). Effect of Brownian motion on thermal conductivity of nanofluids. *Journal of Heat Transfer, 130*(4), 1–13.
12. M. Malika, R. Bhad, & S. S. Sonawane (2021). ANSYS simulation study of a low volume fraction CuO–ZnO/water hybrid nanofluid in a shell and tube heat exchanger. *Journal of the Indian Chemical Society, 98*(11), 100200.
13. W. Yu, & H. Xie (2012). A review on nanofluids: Preparation, stability mechanisms, and applications. *Journal of Nanomaterials, 2012*, 1–17.
14. R. K. Sahu, S. H. Somashekhar, & P. V. Manivannan (2013). Investigation on copper nanofluid obtained through micro electrical discharge machining for dispersion stability and thermal conductivity. *Procedia Engineering, 64*, 946–955.
15. S. S. Sonawane, & M. Malika (2021). Review on CNT based hybrid nanofluids performance in the nano lubricant application. *Journal of Indian Association for Environmental Management (JIAEM), 41*(3), 1–16.
16. M. Malika, & S. S. Sonawane (2021). A comprehensive review on the effect of various ultrasonication parameters on the stability of nanofluid. *Journal of Indian Association for Environmental Management (JIAEM), 41*(4), 19–25.
17. S. Suresh, K. P. Venkitaraj, P. Selvakumar, & M. Chandrasekar (2011). Synthesis of Al$_2$O$_3$-Cu/water hybrid nanofluids using two step method and its thermo physical properties. *Colloids and Surfaces A: Physicochemical and Engineering Aspects, 388*(1–3), 41–48.
18. P. Thakur, S. S. Sonawane, S. Bhaisare, & N. Pandey (2021). Enhancement of pool boiling performance using SWCNT based nanofluids: A sustainable method for the wastewater heat recovery. *Journal of Indian Association for Environmental Management (JIAEM), 41*(4), 7–18.
19. M. Malika, & S. S. Sonawane (2021). The sono-photocatalytic performance of a novel water based Ti+ 4 coated Al (OH) 3-MWCNT's hybrid nanofluid for dye fragmentation. *International Journal of Chemical Reactor Engineering, 19*(9), 901–912.
20. S. Suresh, K. P. Venkitaraj, & P. Selvakumar (2011). Synthesis, characterisation of Al$_2$O$_3$-Cu nano composite powder and water based nanofluids. In *Advanced Materials Research* (Vol. 328, pp. 1560–1567). Trans Tech Publications Ltd.

21. M. Malika, & S. S. Sonawane (2021). Application of RSM and ANN for the prediction and optimization of thermal conductivity ratio of water based Fe_2O_3 coated SiC hybrid nanofluid. *International Communications in Heat and Mass Transfer*, *126*, 105354.

22. A. W. Ahmed, & E. Kalkan (2019). Drilling fluids, types, formation choice and environmental impact. *International Journal of Latest Technology in Engineering Management and Applied Science*, *8*(12), 66–71.

23. H. Jouhara, A. Chauhan, T. Nannou, S. Almahmoud, B. Delpech, & L. C. Wrobel (2017). Heat pipe based systems – Advances and applications, *Energy*, *128*, 729–754.

24. H. Jouhara, N. Khordehgah, S. Almahmoud, B. Delpech, A. Chauhan, & S. A. Tassou (2018). Waste heat recovery technologies and applications, *Thermal Science and Engineering Progress*, *6*, 268–289.

25. B. Delpech, M. Milani, L. Montorsi, D. Boscardin, A. Chauhan, S. Almahmoud, B. Axcell, & H. Jouhara (2018). Energy efficiency enhancement and waste heat recovery in industrial processes by means of the heat pipe technology: Case of the ceramic industry, *Energy*, *158*, 656–665.

26. D. Brough, A. Mezquita, S. Ferrer, C. Segarra, A. Chauhan, S. Almahmoud, N. Khordehgah, L. Ahmad, D. Middleton, & H. I. Sewell (2020). An experimental study and computational validation of waste heat recovery from a lab scale ceramic kiln using a vertical multi-pass heat pipe heat exchanger, *Energy*, *208*, 118325.

27. A. G. Olabi, K. Elsaid, E. T. Sayed, M. S. Mahmoud, T. Wilberforce, R. J. Hassiba, & M. A. Abdelkareem (2021). Application of nanofluids for enhanced waste heat recovery: A review, *Nano Energy*, 84, 105871.

28. M. Rafiq, M. Shafique, A. Azam, & M. Ateeq (2021). Transformer oil-based nanofluid: The application of nanomaterials on thermal, electrical and physico-chemical properties of liquid insulation—A review, *Ain Shams Engineering Journal*, *12*, 555–576.

29. G. Sekrani, & S. Poncet (2018). Ethylene- and propylene-glycol based nanofluids: A litterature review on their thermophysical properties and thermal performances, *Applied Sciences, 8*, 2311.

30. Y. Wang, H. A. I. Al-Saaidi, M. Kong, & J. L. Alvarado (2018). Thermophysical performance of graphene based aqueous nanofluids, *International Journal of Heat and Mass Transfer*, *119*, 408–417.

31. J. J. Fierro, A. Escudero-Atehortua, C. Nieto-Londoño, M. Giraldo, H. Jouhara, & L. C. Wrobel (2020). Evaluation of waste heat recovery technologies for the cement industry, *International Journal of Thermofluids*, *7*, 100040.

32. H. Jouhara, A. Żabnieńska-Góra, N. Khordehgah, D. Ahmad, & T. Lipinski (2020). Latent thermal energy storage technologies and applications: A review, *International Journal of Thermofluids. 5–6*, 100039. doi:10.1016/j.ijft.2020.100039.

33. R. Agathokleous, G. Bianchi, G. Panayiotou, L. Aresti, M. C. Argyrou, G. S. Georgiou, S. A. Tassou, H. Jouhara, S. A. Kalogirou, & G. A. Florides (2019). Waste heat recovery in the EU industry and proposed new technologies, *Energy Procedia*, *161*, 489–496.

34. M. S. Patil, S. C. Kim, J.-H. Seo, & M.-Y. Lee (2016). Review of the thermo-physical properties and performance characteristics of a refrigeration system using refrigerant-based nanofluids, *Energies*, *9*, 22.

35. V. Nair, P. R. Tailor, & A. D. Parekh (2016). Nanorefrigerants: A comprehensive review on its past, present and future, *International Journal of Refrigeration*, *67*, 290–307.

36. S. S. Sanukrishna, & M. J. Prakash (2018). Thermal and rheological characteristics of refrigerant compressor oil with alumina nanoparticles—An experimental investigation, *Powder Technology*, *339*, 119–129.

37. A. Asadi, F. Pourfattah, I. M. Szilágyi, M. Afrand, G. Żyła, H. S. Ahn, S. Wongwises, H. M. Nguyen, A. Arabkoohsar, & O. Mahian (2019). Effect of sonication characteristics on stability, thermophysical properties, and heat transfer of nanofluids: A comprehensive review, *Ultrasonics Sonochemistry*, 58, 104701.

38. I. M. Mahbubul, R. Saidur, & M. A. Amalina (2013). Thermal conductivity, viscosity and density of R141b refrigerant based nanofluid, *Procedia Engineering*, 56, 310–315.

39. M. A. Kedzierski, R. Brignoli, K. T. Quine, & J. S. Brown (2017). Viscosity, density, and thermal conductivity of aluminum oxide and zinc oxide nanolubricants, *International Journal of Refrigeration*, 74, 3–11.

40. O. S. Ohunakin, D. S. Adelekan, T. O. Babarinde, R. O. Leramo, F. I. Abam, & C. D. Diarra (2017). Experimental investigation of TiO_2-, SiO_2- and Al_2O_3-lubricants for a domestic refrigerator system using LPG as working fluid, *Applied Thermal Engineering*, 127, 1469–1477.

41. N. N. M. Zawawi, W. H. Azmi, A. A. M. Redhwan, M. Z. Sharif, & K. V. Sharma (2017). Thermo-physical properties of Al_2O_3-SiO_2/PAG composite nanolubricant for refrigeration system, *International Journal of Refrigeration*, 80, 1–10.

42. S. S. Sanukrishna, & M. J. Prakash (2018). Experimental studies on thermal and rheological behaviour of TiO_2-PAG nanolubricant for refrigeration system, *International Journal of Refrigeration*, 86, 356–372.

43. N. N. M. Zawawi, W. H. Azmi, A. A. M. Redhwan, M. Z. Sharif, & M. Samykano (2018). Experimental investigation on thermo-physical properties of metal oxide composite nanolubricants, *International Journal of Refrigeration*, 89, 11–21.

44. G. Jatinder, O. S. Ohunakin, D. S. Adelekan, O. E. Atiba, A. B. Daniel, J. Singh, & A. A. Atayero (2019). Performance of a domestic refrigerator using selected hydrocarbon working fluids and TiO_2–MO nanolubricant, *Applied Thermal Engineering*, 160, 114004.

45. S. Narayanasarma, & B. T. Kuzhiveli (2019). Evaluation of the properties of POE/SiO2 nanolubricant for an energy-efficient refrigeration system – An experimental assessment, *Powder Technology*, 356, 1029–1044.

46. O. A. Alawi, J. M. Salih, & A. R. Mallah (2019). Thermo-physical properties effectiveness on the coefficient of performance of Al2O3/R141b nano-refrigerant, *International Communications in Heat and Mass Transfer*, 103, 54–61.

47. N. Kumar, & S. S. Sonawane (2016). Experimental study of thermal conductivity and convective heat transfer enhancement using CuO and TiO_2 nanoparticles, *International Communications in Heat and Mass Transfer*, 76, 98–107.

48. A. Adil, T. Baig, F. Jamil, M. Farhan, M. Shehryar, H. M. Ali, & S. Khushnood (2023). Nanoparticle-based cutting fluids in drilling: A recent review, *The International Journal of Advanced Manufacturing Technology*, 1–18.

49. N. Kumar, & S. S. Sonawane (2016). Experimental study of Fe_2O_3/water and Fe_2O_3/ethylene glycol nanofluid heat transfer enhancement in a shell and tube heat exchanger, *International Communications in Heat and Mass Transfer*, 78, 277–284.

50. S. SSonawane, R. S. Khedkar, & K. L. Wasewar (2015). Effect of sonication time on enhancement of effective thermal conductivity of nano TiO_2-water, ethylene glycol and paraffin oil nanofluids and models comparisons, *Journal of Experimental Nanoscience*, 10(4), 310–322.

51. I. M. Mahbubul, A. Saadah, R. Saidur, M. A. Khairul, & A. Kamyar (2015). Thermal performance analysis of Al_2O_3/R-134a nanorefrigerant, *International Journal of Heat and Mass Transfer*, 85, 1034–1040.

52. O. A. Alawi, N. A. C. Sidik, H. W. Xian, T. H. Kean, & S. N. Kazi (2018). Thermal conductivity and viscosity models of metallic oxides nanofluids, *International Journal of Heat and Mass Transfer*, 116, 1314–1325.

53. N. Kumar, S. H. Sonawane, & S. S. Sonawane (2018). Experimental study of thermal conductivity, heat transfer and friction factor of Al_2O_3 based nanofluids, *International Communications in Heat and Mass Transfer*, *90*, 1–10.

54. A. Bhattad, J. Sarkar, & P. Ghosh (2018). Improving the performance of refrigeration systems by using nanofluids: A comprehensive review, *Renewable and Sustainable Energy Reviews*, *82*, 3656–3669.

55. O. A. Alawi, & N. A. C. Sidik (2014). Influence of particle concentration and temperature on the thermophysical properties of $CuO/R134a$ nanorefrigerant, *International Communications in Heat and Mass Transfer*, *58*, 79–84.

56. S. S. Sonawane, & V. Juwar (2016). Optimization of conditions for an enhancement of thermal conductivity and minimization of viscosity of ethylene glycol based Fe_3O_4 nanofluid, *Applied Thermal Engineering*, *109*, 121–129.

57. R. S. Khedkar, S. S. Sonawane, & K. L. Wasewar (2014). Heat transfer study on concentric tube heat exchanger using TiO_2–water-based nanofluid, *International Communications in Heat and Mass Transfer*, 57, 163–169.

58. R. S. Khedkar, S. S. Sonawane, & K. L. Wasewar (2013). Effect of sonication time on Enhancement of effective thermal conductivity nano TiO_2–water, ethelene glycol and paraffin oil nanofluids, *Journal of Experimental Nanosciences*, *10*(4), 310–322.

59. P. Thakur, A. Pargaonkar, & S. S. Sonawane, Introduction to nanofluids. In *Nanofluid Applications for Advanced Thermal Solutions*, Elsevier, 1–19.

60. P. Thakur, & S. S. Sonawane, Synthesis and characterization of nanofluids. In *Nanofluid Applications for Advanced Thermal Solutions*, Elsevier, 20–42.

61. P. Thakur, & S. S. Sonawane (2019). Application of nanofluids in CO_2 capture and extraction from waste water. *Journal of Indian Association for Environmental Management (JIAEM)*, *39*(1–4), 4–8.

62. N. Kumar, & S. S. Sonawane (2016). Influence of CuO and TiO_2 nanoparticles in enhancing the overall heat transfer coefficient and thermal conductivity of water and ethylene glycol based nanofluids, *Research Journal of Chemistry and Environment*, *20*(8), 24–30.

63. V. Juwar, & S. S. Sonawane (2015). Investigations on rheological behaviour of paraffin based Fe_3O_4 nanofluids and its modelling, *Research Journal of Chemistry and Environment*, *19*(12), 16–23.

64. Bashirnezhad, K., Bazri, B., Safaei, M.R., Goodarzi, M., Dahari, M., Mahian, O., Dalkılıça, A.S., & Wongwises, S. (2016). Viscosity of nanofluids: A review of recent experimental studies, *ICHMT*, *73*, 114–123.

65. R. S. Khedkar, S. S. Sonawane, & K. L. Wasewar (2012). Influence of CuO nanoparticles in enhancing the thermal conductivity of water and monoethylene glycol based nanofluids. *International Communications in Heat and Mass Transfer*, *39*(5), 665–669.

66. M. Malika, & S. S. Sonawane (2019). Review on application of nanofluid/nano particle as water disinfectant. *Journal of Indian Association for Environmental Management (JIAEM)*, *39*(1–4), 21–24.

67. W. H. Azmi, K. Abdul Hamid, R. Mamat, & K. V. Sharma (2016). Effects of working temperature on thermo-physical properties and forced convection heat transfer of TiO_2, nanofluids in water ethylene glycol mixture, *Applied Thermal Engineering*, *106*, 1190–1199.

68. T. Ambreen, & M.-H. Kim (2020). Influence of particle size on the effective thermal conductivity of nanofluids: A critical review. *Applied Energy*, *264*, 114684.

69. M. H. Ahmadi, A. Mirlohi, M. AlhuyiNazari, & R. Ghasempour (2018). A review of thermal conductivity of various nanofluids. *Journal of Molecular Liquids*, *265*, 181–188.

70. O. A. Alawi, N. A. C. Sidik, H. W. Xian, T. H. Kean, & S. N. Kazi (2018). Thermal conductivity and viscosity models of metallic oxides nanofluids. *International Journal of Heat and Mass Transfer, 116,* 1314–1325.

71. M. U. Sajid, & H. M. Ali (2018). Thermal conductivity of hybrid nanofluids: A critical review. *International Journal of Heat and Mass Transfer, 126,* 211–234.

72. I. M. Mahbubul (2019). Preparation of nanofluid. In *Preparation, Characterization, Properties and Application of Nanofluid,* W. Andrew, ed., 374, ELSEVIER Publications.

73. I. M. Shahrul, I. M. Mahbubul, S. S. Khaleduzzaman, R. Saidur, & M. F. M. Sabri (2014). A comparative review on the specific heat of nanofluids for energy perspective. *Renewable and Sustainable Energy Reviews, 38,* 88–98.

74. G. M. Moldoveanu, & A. A. Minea (2019). Specific heat experimental tests of simple and hybrid oxide-water nanofluids: Proposing new correlation. *Journal of Molecular Liquids, 279,* 299–305.

75. S. M. S. Murshed, & P. Estellé (2017). A state of the art review on viscosity of nanofluids. *Renewable and Sustainable Energy Reviews, 76,* 1134–1152.

76. D. Chavan, & A. Pise (2019). Experimental investigation of effective viscosity and density of nanofluids. *Materials Today: Proceedings, 16,* 504–515.

77. A. Mariano, M. J. Pastoriza-Gallego, L. Lugo, A. Camacho, S. Canzonieri, & M. M. Piñeiro (2013). Thermal conductivity, rheological behaviour and density of non-Newtonian ethylene glycol-based SnO_2 nanofluids. *Fluid Phase Equilibria, 337,* 119–124.

78. Y. Xuan, Q. Li, & P. Tie (2013). The effect of surfactants on heat transfer feature of nanofluids. *Experimental Thermal and Fluid Science, 46,* 259–262.

79. N. Ali, J. A. Teixeira, & A. Addali (2018). A review on nanofluids: Fabrication, stability, and thermophysical properties. *Journal of Nanomaterials, 2018,* 1–33.

80. X. J. Wang, & D. S. Zhu (2009). Investigation of pH and SDBS on enhancement of thermal conductivity in nanofluids. *Chemical Physics Letters, 470*(1–3), 107–111.

81. S. A. Angayarkanni, & J. Philip (2014). Effect of nanoparticles aggregation on thermal and electrical conductivities of nanofluids. *Journal of Nanofluids, 3*(1), 17–25.

82. I. M. Mahbubul, E. B. Elcioglu, R. Saidur, & M. A. Amalina (2017). Optimization of ultrasonication period for better dispersion and stability of TiO_2–water nanofluid. *Ultrasonics Sonochemistry, 37,* 360–367.

83. Paula I.P. Soares, César A.T. Laia, A. Carvalho, Laura C.J. Pereira, Joana T. Coutinho, Isabel M.M. Ferreira, Novo, Carlos M.M. & João Paulo Borges (2016). Iron oxide nanoparticles stabilized with a bilayer of oleic acid for magnetic hyperthermia and MRI applications. *Applied Surface Science, 383,* 240–247.

84. S. Umar, F. Sulaiman, N. Abdullah, & S. N. Mohamad (2018, November). Investigation of the effect of pH adjustment on the stability of nanofluid. In *AIP Conference Proceedings* (Vol. 2031, No. 1, p. 020031). AIP Publishing LLC.

85. S. K. Sharma, & S. M. Gupta (2016). Preparation and evaluation of stable nanofluids for heat transfer application: A review. *Experimental Thermal and Fluid Science, 79,* 202–212.

86. D. Li, B. Hong, W. Fang, Y. Guo, & R. Lin (2010). Preparation of well-dispersed silver nanoparticles for oil-based nanofluids. *Industrial & Engineering Chemistry Research, 49*(4), 1697–1702.

87. C. Rodrıguez-Abreu (2016). Nanocolloids: Some basic concepts and principles of their stabilization. *Nanocolloids, 1.* ISBN number 978-0-12-801578-0.

88. Y. T. He, J. Wan, & T. Tokunaga (2008). Kinetic stability of hematite nanoparticles: The effect of particle sizes. *Journal of Nanoparticle Research, 10*(2), 321–332.

89. D. A. Walker, B. Kowalczyk, M. O. de La Cruz, & B. A. Grzybowski (2011). Electrostatics at the nanoscale. *Nanoscale*, *3*(4), 1316–1344.
90. H. J. Kim, S. H. Lee, J. H. Lee, & S. P. Jang (2015). Effect of particle shape on suspension stability and thermal conductivities of water-based bohemite alumina nanofluids. *Energy*, *90*, 1290–1297.
91. T. Vítěz, & P. Trávníček (2014). Study of settling velocity of sand particles located in wastewater treatment plant. *Acta Universitatis Agriculturae et Silviculturae Mendelianae Brunensis*, *59*, 28.
92. S. Chakraborty, I. Sarkar, K. Haldar, S. K. Pal, & S. Chakraborty (2015). Synthesis of Cu–Al layered double hydroxide nanofluid and characterization of its thermal properties. *Applied Clay Science*, *107*, 98–108.
93. S. Chakraborty, I. Sarkar, A. Ashok, I. Sengupta, S. K. Pal, & S. Chakraborty (2018). Thermo-physical properties of Cu-Zn-Al LDH nanofluid and its application in spray cooling. *Applied Thermal Engineering*, *141*, 339–351.
94. K. S. Hong, T. K. Hong, & H. S. Yang (2006). Thermal conductivity of Fe nanofluids depending on the cluster size of nanoparticles. *Applied Physics Letters*, *88*(3), 031901.
95. S. Chakraborty, & P. K. Panigrahi (2020). Stability of nanofluid: A review. *Applied Thermal Engineering*, *174*, 115259.
96. S. J. Aravind, P. Baskar, T. T. Baby, R. K. Sabareesh, S. Das, & S. Ramaprabhu (2011). Investigation of structural stability, dispersion, viscosity, and conductive heat transfer properties of functionalized carbon nanotube based nanofluids. *The Journal of Physical Chemistry C*, *115*(34), 16737–16744.
97. I. M. Mahbubul (2019). Stability and dispersion characterization of nanofluid. *Preparation, Characterization, Properties and Application of Nanofluid*, 47–112. Elsevier publication. ISBN NO: 9780128132456.
98. A. K. Singh, & V. S. Raykar (2008). Microwave synthesis of silver nanofluids with polyvinylpyrrolidone (PVP) and their transport properties. *Colloid and Polymer Science*, *286*(14), 1667–1673.
99. D. F. Swinehart (1962). The Beer-Lambert law. *Journal of Chemical Education*, *39*(7), 333.
100. W. Chamsa-Ard, S. Brundavanam, C. C. Fung, D. Fawcett, & G. Poinern (2017). Nanofluid types, their synthesis, properties and incorporation in direct solar thermal collectors: A review. *Nanomaterials*, *7*(6), 131.
101. L. Fedele, L. Colla, S. Bobbo, S. Barison, & F. Agresti (2011). Experimental stability analysis of different water-based nanofluids. *Nanoscale Research Letters*, *6*(1), 1–8.

2 Application of Nanofluids for the Solar Collectors

2.1 INTRODUCTION

Over the past few years, the use of solar energy has increased. Petroleum derivatives are conventional, and their availability decreasing day by day. The price of petroleum is also increasing day by day. The flash increase in world population and growth of industries caused an energy crisis. The rapid advancement of socio-economic needs is also the reason behind the global energy crisis. The use of traditional petroleum derivatives increased while the availability of these products decreased day by day. Our planet Earth gets very much more energy in just 1 hour from the sun than that consumed by the whole world in 1 year. Now people have started focusing on renewable energy resources like solar, wind, geothermal, hydro, etc.

As we have seen in the previous chapter, solids have comparatively more thermal conductivity than liquids. Thus, the suspension of nanoparticles (generally of metal oxide and carbon nanotubes) and base fluid have more thermal conductivity than the base fluid alone. Nanofluids are nothing but the engineered suspension of the nanoparticles of metal oxides and base fluids. Nanofluids exhibit better thermal properties than the base fluid. They possess better heat transfer coefficient and thermal conductivity. These properties are useful for performing better heat exchange and boiling operations. Thus, the researchers are working to develop a highly efficient nanofluids system to use in applications like solar panels [1]. Gupta et al. [2] have compared the performance of various nanofluids based on their thermo-physical property change. Various parameters important for thermal conductivity are represented in Figure 2.1. The performance of the nanofluids in the car radiator and solar panel is also discussed in detail [3]. Ali et al. [4] also suggested the various nanofluids with high efficiency of heat transfer. They highlighted the important role of flow patterns and geometry of the heat exchange operation. Ali et al. [5] highlighted the issues faced by the researchers in using the nanofluids on a larger scale. The main issues highlighted by the author are the stability of nanofluids and, the synthesis cost of highly efficient nanoparticles like CNT.

Enhancement of heat transfer is the main factor energy energy-saving and compact designs. Most solar water heating Systems contain two main parts: solar collector and storage tank. Flat plate solar collector is the most common solar collector but its efficiency is relatively low. The efficiency of the collector depends not only on how much the absorber captures solar energy but also on how much heat is transferred to the working fluid. To avoid these drawbacks, for solar thermal

DOI: 10.1201/9781003404767-2

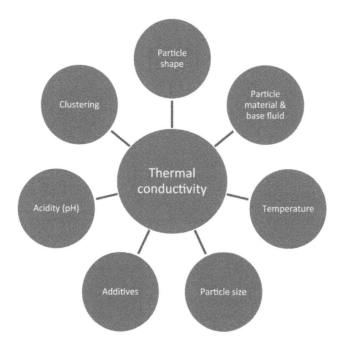

FIGURE 2.1 Different important parameters affecting the thermal conductivity [2].

utilization direct solar absorption collector is being used. In direct solar absorption collector sunlight is absorbed directly from the working fluid and is then exported in the form of heat. Here black nanofluid suspensions are limited because of severe abrasion, plug problems of coarse particles and sedimentation.

As we have seen in the first chapter, metal oxide-based nanoparticles are extensively used in nanofluids synthesis. Said et al. [6] used various metal oxide-based nanoparticles for the solar collector study. The refractive index of the synthesized nanofluids and water are shown in Figure 2.2. The refractive index

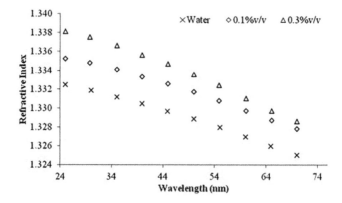

FIGURE 2.2 Change in the refractive index at different temperatures [6].

decreases with the increase in the wavelength of solar energy, i.e., at high temperatures. Ali et al. [7] conducted experiments with the TiO_2 nanoparticles and reported the high efficiency of the TiO_2-based nanofluids. Synthesis methods used for the nanofluids play a vital role in the thermo-physical properties of the nanofluids [8]. More than a 40% increase in the Nusselt number is also observed by the authors in the mini channel-based geometry [9]. The geometry of the system is a very important parameter for the efficiency of the nanofluids. Thus, it is difficult to develop the generalized formulas of the thermo-physical properties of the nanofluids [10].

Said et al. [11] conducted several experiments to understand the usefulness of nanofluids in the effective utilization of solar energy. They have used TiO_2-based nanofluids for the solar panel study. The author highlighted that optical properties are also important for the application of nanofluids in the solar panel study [12]. The optical properties are also studied for the SWCNT-based (single walled carbon nanotubes) nanofluids. For 0.25 vol% of nanoparticle concentration in the nanofluids, the extinction coefficient reported is 8.2 cm^{-1} and for 0.1 vol% concentration the extinction coefficient is 4.98 cm^{-1} [13]. Base fluids generally used in the reviewed study are water, ethylene glycol, a mixture of water and ethylene glycol, propylene glycol, ethanol, glycol, therminol VP-1, thermal oil, paraffin, ionic liquid, etc. Nanofluid stability depends on mainly temperature during the preparation of the sample and the sedimentation phenomenon is inversely proportional to temperature during magnetic stirrer mixing.

Various types of solar collectors are used to absorb the solar energy into the working fluid. This absorbed energy is then transferred to further application [14]. Solar energy is very important and available to a large extent. But, we use a very limited amount of energy among this available energy. Solar collectors are more widely used than any other application of solar energy [15]. Generally, water is used as a working fluid in the solar collector and recent studies in the nanofluids have proved that nanofluids have much better performance than water as a working fluid. Various types of collectors are shown in Figure 2.3 [16]. The American Society of Heating, Refrigerating and Air-conditioning Engineers (ASHRAE) has developed a few guidelines to test solar collector efficiency. The efficiency of the solar collector is measured in terms of energy efficiency and exergy efficiency. In this analysis, it is assumed that the energy is made up of exergy and anergy. Exergy is the measure of a useful amount of energy and anergy is the remaining amount of energy that cannot be used due to thermodynamic restrains. Abid et al. [17] used alumina-based nanofluids for the exergy analysis of the 60 MWe steam power plant. This plant consists of the parabolic trough solar collectors. Thermal oil is used as a working fluid for this study. The energy efficiency of this system is reported as 22.64% and the exergy efficiency of this system is 23.83%. Abid et al. [18] also performed experiments with the solar parabolic dish-assisted solar collector and it is observed that the energy efficiency of this setup is 34.5% and exergy efficiency is 37.1%. Ali et al. [19] compared the performance of various geometries of the parabolic trough collector and a maximum of 2.71% enhancement in the thermal performance was observed. Ali et al. [20] used a supercritical Brayton cycle-based solar collector for the study and a 33.73% enhancement in energy efficiency and a 36.27% increase in exergy efficiency are reported. Khan et al. [21] used silica as a nanoparticle and

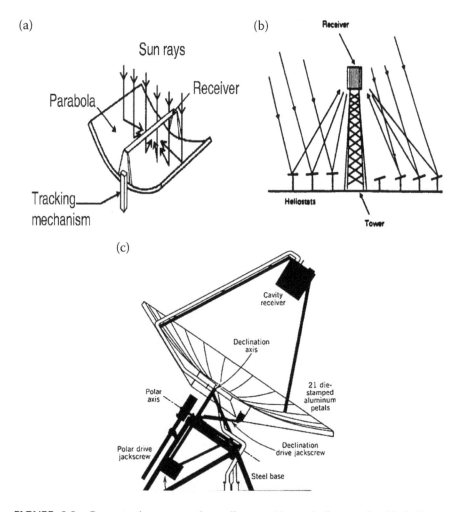

FIGURE 2.3 Concentrating type solar collectors (a) parabolic trough, (b) heliostat, (c) parabolic dish [16].

VP1 as a working fluid for the solar panels. The study showed the maximum energy efficiency is nearly 39% and the exergy efficiency is nearly 42%.

2.2 TYPES OF SOLAR COLLECTORS

2.2.1 FLAT PLATE COLLECTORS

Flat plate solar collectors are the simplest type of solar collector. These types of collectors are cheaper than other solar collectors and easy to fabricate and install. However, these collectors have comparatively less efficiency than other solar collectors [22]. Figure 2.4 represents the schematic representation of a flat plate solar collector. Glass sheets, Back cover, header pipes, tubes are parts of a flat plate solar collector. The glass used in this collector should be clean and optically

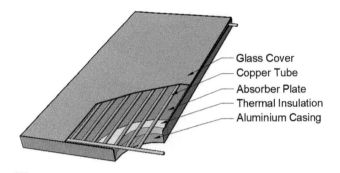

Glass Cover
Copper Tube
Absorber Plate
Thermal Insulation
Aluminium Casing

FIGURE 2.4   Schematic representation of flat plate solar collector [33].

transparent. Nearly 90% of sunlight passes through the glass. The glass prevents the backscattering of the sunlight from the working fluid [23]. The channel for the working fluid is important for the transportation of working fluid. The back cover is used to prevent the transmittance of solar energy from the channel. Flat plate solar collectors are easy to fabricate. The researchers extensively use this type of solar collector to study the performance of various nanofluid systems in this type of collector [24].

Yousefi et al. [25–29] conducted experiments with multi-walled carbon nanotubes MWCNT/water nanofluids system and reported the effect of pH on the nanofluids. The author used Triton X-100 surfactant for the stability of nanofluids. In another study, the Author used alumina and a 28% increase in efficiency was reported. Ali et al. [30] used CuO-based nanofluids for the solar panel study and observed an increased solar efficiency of 16.7%. Said et al. [31] conducted experiments with the Al_2O_3/water system and a nearly 83% increase in the efficiency was reported. In another study, Said et al. [32] observed nearly a 4% entropy decline by use of SWCNT/water nanofluids.

2.2.2 Photovoltaic Thermal Collectors

Photovoltaic thermal (PVT) cells are used to produce electricity and absorb the solar energy into the working fluid. Figure 2.5 represents the PVT cells mechanism. PV cells are attached to the top of serpentine tubes of working fluid. Generally, 8–15% of incident solar energy is used for electricity production and the remaining energy is absorbed in the working fluid for further application [34]. This type of solar collector is comparatively costly to install but easy to maintain and thus they are preferred in many cases. ZnO/water and silicon carbide/water systems have proved to be one of the best nanofluids systems for the PVT cell application [35]. Al-waeli et al. [36] conducted experiments with the Silicon carbide/water nanofluids and reported that, at 3 wt% of nanoparticle, the thermal efficiency can be increased up to 100%. The electric efficiency is observed as 24%. Thus, the overall efficiency of the nanofluids is reported as 89%. In another study, Al-waeli [37] conducted experiments with various nanoparticles like alumina, CuO, and silicon carbide and at 4 wt% the performance of silicon carbide-based nanofluids is best. Sardarabadi et al. [38] also reported a 3.6% increase in the thermal efficiency at

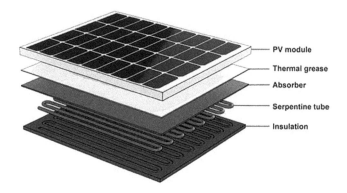

FIGURE 2.5 Schematic representation of PVT collectors [40].

3 wt% of SiC/water nanofluids and a 24.31% increase in the exergy efficiency. In another study, Sardarabadi et al. [39] concluded that TiO_2 and ZnO have better performance for PVT cells than the SiC-based nanofluids.

2.2.3 EVACUATED TUBE SOLAR COLLECTOR

These types of solar collectors are designed in such a way that they operate at vacuum conditions and thus they can operate at high temperature conditions. Figure 2.6 represents the various components of the vacuum/evacuated type of solar collector. The vacuum condition is useful for the prevention of thermal loss [41]. Due to the vacuum condition, the efficiency of these solar collectors is better than the flat plate solar collector, but the cost of these collectors is high. Tubes used in these collectors are heat pipes or U-shaped tubes. The performance of heat pipes is better than U-shaped tubes but in the rainy season, U-shaped collectors have better efficiency [42]. Researchers have conducted the study of various nanofluids in this collector and enhancement is observed for each case [43].

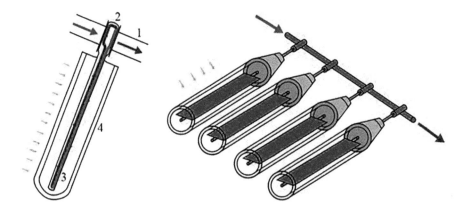

FIGURE 2.6 Schematic representation of the evaluated tube solar collectors with important parts. 1—manifold header, 2—heat pipe head, 3—heat pipe, and 4—vacuum glass tube [46].

Sabiha et al. [44] conducted the experiments with the 'single walled carbon nanotubes' SWCNT-based nanofluids and a nearly 93% increase in the thermal efficiency was observed at 0.2 wt% nanoparticle loading. Park and Kim [45] conducted experiments with the MWCNT-based nanofluids and an increase in the thermal efficiency of 12.6% was observed.

2.2.4   DIRECT ABSORPTION SOLAR COLLECTORS

Direct absorption solar collectors are the easiest type of solar collector to fabricate. The working fluid is allowed to pass through the channel and the upper side of the channel is closed by the glass cover [47]. Incident radiation of the sun is directly captured by the working fluid. The exergy is lost only at three stages. Firstly at the glass cover, as the incident radiation falls on the glass cover. Some energy gets reflected and thus the energy gets lost secondly, some energy gets absorbed by the glass cover, and the remaining energy gets transmitted through the glass cover. Some amount of energy is again lost in the transmittance and absorbance by the working fluid. The absorbed energy passes with the working fluid and the working fluid is set for further application [48]. Researchers have conducted experiments with the various nanofluids systems and reported that similar results as a flat plate solar collector are observed. Researchers like Tyagi et al. [49] reported that the efficiency of the direct absorption solar collector is better than the flat plate solar collector by 10% [50].

2.3   ENERGY ANALYSIS

The most basic formula required for the energy analysis is the energy balance of the system. Equation 2.1 is the energy balance equation for all types of solar collectors [51]. Here, Q_u is the amount of energy absorbed by the solar collector (W). m is the mass flow rate of working fluid (kg/sec). Cp is the specific heat of working fluid (J/kg °C) and T_{in} and T_{out} are the temperatures at the inlet of the Solar collector (°C) and outlet of the solar collector.

$$Q_u = mCp\,(T_{out} - T_{in}) \tag{2.1}$$

2.3.1   FLAT PLATE COLLECTOR

The efficiency of a solar collector can easily calculated by the ratio of energy absorbed by the solar collector to energy available for the solar collector. The calculation uses Equation 2.2 [52]. Here, A_c is the area available for the solar collector (m^2) and I_T is incident radiation (W/m^2)

$$\eta = \frac{Q_u}{A_c I_T} \tag{2.2}$$

From Equation 2.1, we know the value of absorbed energy by the solar collector. We get

$$\eta = \frac{mCp\,(T_{out} - T_{in})}{A_c I_T} \tag{2.3}$$

Another way to calculate the thermal efficiency of solar panels is to subtract the thermal losses from the absorbed energy parameter. Equation 2.4 is used to calculate the absorbed energy by subtracting the lost energy parameter from the available energy parameter. F_R is a friction factor $(\tau\alpha)$ is a product of the transmittance-absorbance parameter U_L is a lost energy parameter. Solar energy collected by the solar collector can be calculated by subtracting thermal losses from the absorbed thermal radiation from the sun. i.e., represented by Equation 2.4.

$$\text{i.e.,}\quad Q_u = A_C F_R [I_T\,(\tau\alpha) - U_L\,(T_i - T_a)] \tag{2.4}$$

Comparing Equations 2.4 and 2.3, we get,

$$\eta = \frac{mCp\,(T_{out,f} - T_{in,f})}{A_c I_T} = F_R\,(\tau\alpha) - U_L F_R \left(\frac{T_i - T_a}{I_T}\right) \tag{2.5}$$

Equation 2.5 is used to compare the performance of various nanofluid systems. Efficiency value is calculated from Equation 2.3 and plotted on the Y-axis and $(T_i - T_a)/I_T$ values are plotted on the X-axis [53]. This will give us the linear plots for different temperatures and incident radiation. These lines will have a negative slope. There will be no reading with zero value in this plot. This is because the inlet temperature will always be more than the ambient temperature. The intersection of these lines on the Y-axis is the value of $F_R\,(\tau\alpha)$ [54].

2.3.2 PHTOVOLTAIC-THERMAL SOLAR COLLECTOR

The total efficiency of the PVT system is the summation of efficiency of thermal efficiency and electric efficiency. As shown in Equation 2.6

$$\eta_{PVT} = \eta + \eta_{el} \tag{2.6}$$

η is calculated in the same way as we calculated for the flat plate solar collector. The important formula for the PVT system efficiency calculation is the electric output efficiency (η_{el}). This efficiency is calculated by Equation 2.7.

$$\eta_{el} = \eta_{ref} \cdot [1 - \beta \cdot (T_{PV} - T_{ref})] \tag{2.7}$$

In the above equation, β is the photovoltaic coefficient dependent on the temperature η_{ref} is electric efficiency at the reference temperature and T_{PV} is the operating temperature of the photovoltaic cell.

2.3.3 THE EVACUATED TUBE TYPE SYSTEM

The strategy used to calculate the flat plate solar collector is also useful for the calculation of the efficiency of the evacuated type solar collector [55].

2.3.4 DIRECT ABSORPTION SOLAR COLLECTOR

Generally, Equations 2.1–2.5 are used to calculate the energy efficiency of nanofluids in direct absorption solar collectors [56]. The pressure drop is important for the calculation of the efficiency of nanofluids in the direct absorption solar collector. Thus, the basic fluid mechanics equations like Darcy Weis batch equations are used to determine the pressure drop across the solar collector [57].

2.4 PARAMETERS AFFECTING THE SOLAR PANEL PERFORMANCE

Sahin et al. [58] reviewed the various parameters of solar collectors and emphasized that the correct combination of particle size, adequate dispersion, and pH value of nanofluid helps in the improvement of efficiency. When compared with other nanoparticles, carbon nanotubes results in larger enhancement.

Tiwari et al. [59] theoretically investigated the efficiency of flat plate solar collectors with Al_2O_3/water nanofluids. Amalraj and Michael [60] synthesized oxide ceramics of CuO and Al_2O_3 by a step combustion process using glycine as an organic fuel. The FT-IR spectra of the CuO and alumina nanoparticles are shown in Figure 2.7. Acquired nanoparticles are characterized by powder X-ray diffractometer. The X-ray diffraction of alumina is shown in Figure 2.8. It confirms CuO with monoclinic and Al_2O_3 with rhombohedral phase, respectively. Scanning electron microscopy analysis shows CuO as spherical shaped and Al_2O_3 as leaf shaped. These results are represented in Figure 2.9 and EDAX results are shown in Figure 2.10. Walke et al. [61] studied the various nanofluids and their application for the solar collector.

Sajid and Bicer [62] reviewed various nanofluids and showed that transmittance is different for different nanofluids. Petko [63] studied the effect of carbon nanoparticles concentration in propylene glycol and water-based nanofluids on the solar energy absorbance. These nanofluids are obtained by primary treatment of used particles with

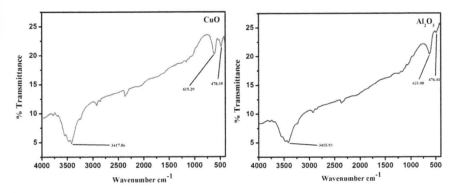

FIGURE 2.7   FT-IR results of CuO and alumina nanoparticles [60].

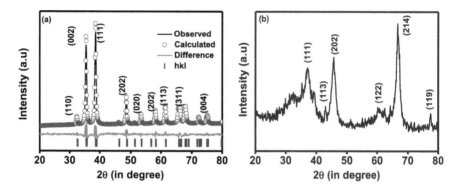

FIGURE 2.8 X-ray diffraction results of (a) CuO nanoparticle and (b) alumina nanoparticle [60].

FIGURE 2.9 SEM results of (a and b) CuO nanoparticle and (c and d) alumina nanoparticle [60].

H_2O_2. They found the optimal concentration for both dispersion media was 0.2 g dm^{-3}. Extinction coefficients are slightly higher than propylene glycol for water-based fluids. Extinction coefficients decrease with wavelength. Transmittance is calculated by the Bourguer Lambert-Beer law

$$T = \frac{I}{Io} = 10^{-\alpha l} = 10^{-\varepsilon l C} \qquad (2.8)$$

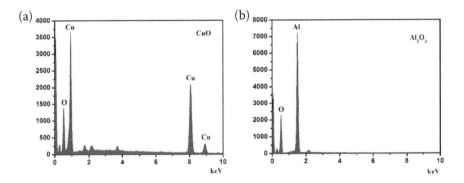

FIGURE 2.10 Results of (a) CuO nanoparticles and (b) alumina nanoparticles [60].

where T = transmittance, Io and I are the intensity of the incident and transmitted light, l = length path, α = absorption coefficients, ε = molar specific extinction coefficient, and C = concentration.

Chen [64] investigated the stability, thermal conductivity, and optical properties of SiC-saline water nanofluid. Figure 2.11 represents the TEM images of SiC nanoparticles. Experimental results have shown that SiC increases by more than 6% thermal conductivity at 0.4 vol% SiC nanofluids. This result is shown in Figure 2.12 and less than 1% luminousness at 0.4 vol% SiC nanofluid. These results are shown in Figure 2.13.

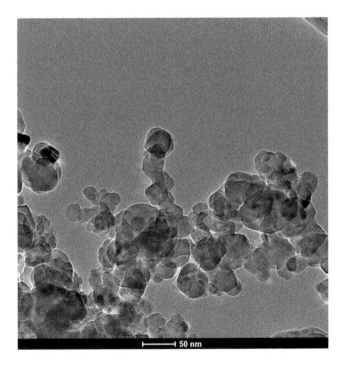

FIGURE 2.11 TEM images of SiC nanoparticle [64].

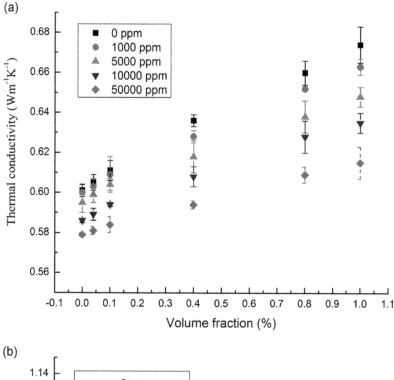

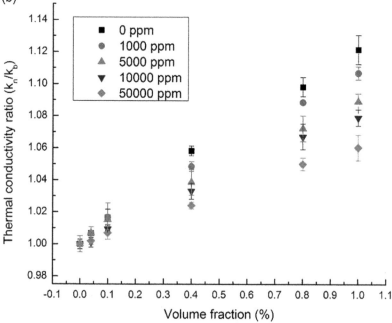

FIGURE 2.12 (a) Thermal conductivity vs. volume fraction for the SiC nanofluid and (b) thermal conductivity ratio vs. volume fraction for SiC nanofluids [64].

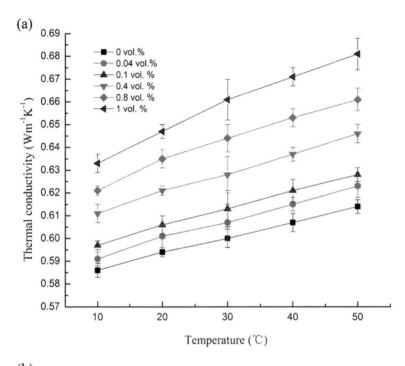

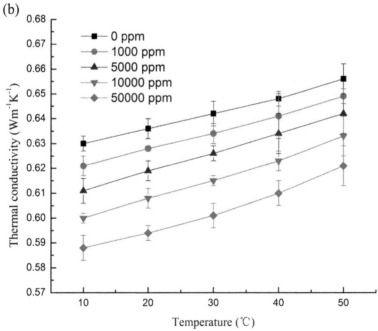

FIGURE 2.13 Temperature dependency of various concentrations of nanofluids and various concentrations of salinity [64].

Wanga et al. [65] suggest an application of Chinese ink as nanofluids in solar thermal conversion. The intensity of the dispersion of nanofluids was characterized by a spectrophotometer. Compared to the reported Cu and CuO nanofluids, those Chinese inks showed satisfactory stability of dispersion. The results obtained revealed that Chinese ink nanofluid has a surprisingly high picture-thermal transformation efficiency than Cu and CuO nanoparticles. Carbon black is determined by Chinese ink nanofluid to convert the image of the sun to heat. The bone attached to the Chinese face ink nanoparticle also plays an important role in preventing carbon black nanoparticles from synthesizing and placement. The process of preparing Chinese ink is quite simple and the materials are cheap as well as widely available, promising extensive industrial use of Chinese solar thermal ink.

Yang et al. [66] discovered heat temperatures, temperature cycles, and heating time were all very influential stability of nanofluids. Nanofluids tend to mix quickly at medium temperatures or after several temperature cycles even if they have a high hardness at room temperature. According to our orthogonal experiment, the main factors affecting the stability of thermal impact were the time of periodic regeneration, particle size, particle concentration, and type of dispersion. The thermal conductivity of CuO/nanofluids of oil with 0.2% volume fraction was 3.8% higher than that of pure oil base fluid, making its active energy in the solar collector gain 67.6%

Sahin et al. [67] did experiments with Al_2O_3, TiO_2, SiO_2, and Cu nanoparticles in water as a fluid. They found that Cu-water produces the lowest entropy generation, Al_2O_3-water shows the highest heat transfer coefficient and SiO_2-water the lowest. Figure 2.14 represents the results reported by the authors. The comparative results are tabulated in the Table 2.1.

Li et al. [86] demonstrated a straightforward model for broad-band solar thermal nanofluids. CuO has high absorption in the visible region and ATO nanofluids have high absorption near the infrared region. So, the combination of both nanofluids worked for the visible and infrared regions.

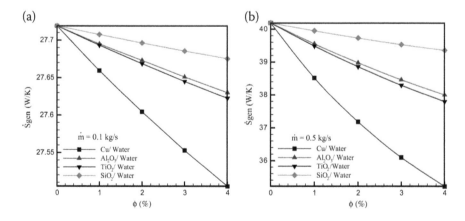

FIGURE 2.14 Variation of entropy generation for different nanofluids at (a) 0.1 kg/sec mass flow rate and (b) 0.5 kg/sec mass flow rate [67].

TABLE 2.1

A Literature Review of Nanofluids-Based Solar Collector Studies

References	Collector Type	Nanofluids Type	Results
Tiwari et al. [59]	Flat-plate	Al_2O_3/water	Using Al_2O_3 increases thermal efficiency up to 31.64%
Amalraj and Michael [60]	Solar cooling panel	CuO and Al_2O_3	Cooling efficiency is 18.2%
Petko [63]	Direct absorption solar collector	H_2O_2-treated C particles	Solar absorption in the wavelength range 200–2500 nm is over 96% for optical C concentration 0.2 g dm^{-3}. Photothermal properties of C black nanofluids showed a good temperature growth rate with solar irradiation time.
Navas [68]	Direct absorption solar collector	Graphene oxide-based nanofluid	Increases thermal efficiency
Cheng et al. [69]	Direct absorption solar collector	Fe_3O_4/ionic liquid	Enhancement of optical properties, transmittance, scattering, and extinction coefficient studied.
Chen et al. [64]	Solar distillation system	SiC nanofluids with saline water	Performance of nanofluids with saline water studied
Huang et al. [70]	Solar photo-thermal conversion	Au nanofluids	Experimental and theoretical investigation done for solar photo-thermal conversion of Au nanofluids.
Zuo et al. [71]	Solar desalination system	Carbon nanotube nanofluids	0.04 wt% MWCNT nanofluids absorbed almost 100% solar energy for more than 1-cm thickness of fluid. Evaporation efficiency enhanced from 24.91% (0 wt%) to 76.65% (0.04 wt%). MWCNT concentration is in weight %.
Favale et al. [72]	Solar energy systems	Al_2O_3-diathermic oil	Viscosity increases with volume concentration and cluster size. Non-Newtonian behaviour of nanofluids with and without surfactants.
Ehsan [73]	Solar heat exchanger	TiO_2 nanofluids	Maximum 21% increment in average heat transfer coefficient reached by using TiO_2-water nanofluids at 2.3% volume concentration.
Behshad [74]	Heat storage system	–	Experimental investigation of storage of heat by using phase change material.

(Continued)

TABLE 2.1 (Continued)
A Literature Review of Nanofluids-Based Solar Collector Studies

References	Collector Type	Nanofluids Type	Results
He et al. [75]	Solar thermal energy system	Cu-H_2O nanofluids	Investigated transmittance influencing factors like particle size, optical path, and mass fraction and also measured extinction coefficients
Hu et al. [76]	–	Salt-based SiO_2 nanofluids	Freeze drying method was suggested to formulate nanofluids Shah London correlation tested.
Kashani et al. [77]	Solar energy absorption	Graphite nanofluids	Developed a method to solve transport equations.
Mahian et al. [67]	Mini-channel-based solar collector	–	Some recommendations are provided for future studies.
Rajan and Manikandan [78]	Solar energy collection	Sand-propylene, glycol-water	Did experiments to study viscosity, non-covalent interactions, etc.
Taylor et al. [79]	Solar thermal collectors	Carbon nanotube	Investigated thermal stability of plasma and multi-walled carbon nanotubes dispersed in different base fluids.
Rativa and Gómez-Malagón [80]	Solar radiation absorption	Metallic nano-ellipsoids	Used Maxwell Garnet model.
Sani et al. [81]	Solar energy applications	Graphite/diamond, ethylene glycol	Investigated optical properties, also proved the creation of vapor bubbles in the base fluid by limiting effects.
Sheikh et al. [82]	Solar collectors	–	They presented the Caputo-Farizio and Atangna-Baleanu fractional model. They showed that by adding aluminium oxide efficiency can be enhanced by 5.2%.
Taylor et al. [83]	Solar collector	–	They presented a nanofluid receiver's design.
Sani et al. [84]	Solar energy applications	Graphene	Analyzed two types of nanofluid, consisting of polycarboxylate graphene nanoplatelets and sulphonic acid functionalized graphene.
Tiwari et al. [85]	Solar collector	MgO/water nanofluid	They analyzed the performance based on the first law of energy balance and the qualitative nature of the energy flow rate.

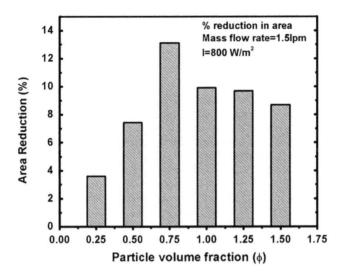

FIGURE 2.15   Exergy efficiency of MgO/water nanofluids and water [85].

Rativa and Gómez-Malagón [80] used gold and silver nano ellipsoids to study linear optical absorption in the visible and near-infrared region using the Maxwell garnet model. They show that solar solar-weighted absorption coefficient near to ideal solar absorber condition can be calculated by tuning NEs geometry.

Mustafa [87] presented a paper on the enhancement of the performance of solar collectors based on alumina water nanofluids. He reported that 85.63% thermal efficiency was achieved by using a non-adiabatic bottom panel by adding aluminium in pure water. An efficiency of 100% was attained with lesser base temperatures in higher nanoparticle volume function. The reduction in the area to achieve the same efficiency is also observed. These results are shown in Figure 2.15.

Vallejo et al. [84] used two different polycarboxylate nano-suspensions chemically modified graphene nanoplatelets (P-GnP) and sulfonic acidapidated graphene nanoplatelets (S-GnP) dispersed in the tested water, in three different locations (0.005 wt%, 0.025 wt% and 0.05 wt%). Dispersion, and especially P-GnP, has shown satisfactory long-term stability. Spectrophotometric measurements were allowed to detect the coefficients of spectral extraction and to test the potential for direct solar absorber applications. In both types of colloids, the sun's rays are almost entirely at a distance of 5 to 20 mm, with differences between samples attached to concentrations and different performances of nano additives.

2.5   CONCLUSION

This chapter gives a review of the latest studies regarding the nanofluids application for effective solar energy utilization. In this chapter, various types of solar collectors are discussed in detail and the recent advances in this regard are reported. Various parameters affecting the efficiencies of different types of solar collectors are also reported in the chapter. Only a few correlations are available regarding transmittance and extinct numbers. More systematic experimental studies can be conducted to develop effective nanofluid systems for solar collector applications.

REFERENCES

1. A. Wahab, A. Hassan, M. A. Qasim, H. M. Ali, H. Babar, & M. U. Sajid (2019). Solar energy systems – Potential of nanofluids. *Journal of Molecular Liquids*, *289*, 111049.
2. M. Gupta, V. Singh, R. Kumar, & Z. Said (2017). A review on thermophysical properties of nanofluids and heat transfer applications. *Renewable and Sustainable Energy Reviews*, *74*, 638–670.
3. Z. Said, M. E. H. Assad, A. A. Hachicha, E. Bellos, M. A. Abdelkareem, D. Z. Alazaizeh, & B. A. Yousef (2019). Enhancing the performance of automotive radiators using nanofluids. *Renewable and Sustainable Energy Reviews*, *112*, 183–194.
4. M. U. Sajid, & H. M. Ali (2019). Recent advances in application of nanofluids in heat transfer devices: A critical review. *Renewable and Sustainable Energy Reviews*, *103*, 556–592.
5. S. Javed, H. M. Ali, H. Babar, M. S. Khan, M. M. Janjua, & M. A. Bashir (2020). Internal convective heat transfer of nanofluids in different flow regimes: A comprehensive review. *Physica A: Statistical Mechanics and its Applications*, *538*, 122783.
6. Z. Said, R. Saidur, & N. Rahim (2014). A optical properties of metal oxides based nanofluids. *International Communications in Heat and Mass Transfer*, *59*, 46–54.
7. H. M. Ali, M. U. Sajid, & A. Arshad (2017). Heat transfer applications of TiO_2 nanofluids. Application of titanium dioxide. *InTech, Rijekam*, Ch. 9.
8. H. M. Ali, H. Babar, T. R. Shah, M. U. Sajid, M. A. Qasim, & S. Javed (2018). Preparation techniques of TiO_2 nanofluids and challenges: A review. *Applied Sciences*, *8*(4), 587.
9. M. U. Sajid, H. M. Ali, A. Sufyan, D. Rashid, S. U. Zahid, & W. U. Rehman (2019). Experimental investigation of TiO_2-water nanofluid flow and heat transfer inside wavy mini-channel heat sinks. *Journal of Thermal Analysis and Calorimetry*, *137*(4), 1279–1294.
10. M. E. M. Soudagar, M. A. Kalam, M. U. Sajid, A. Afzal, N. R. Banapurmath, N. Akram, & A. Saleel C (2020). Thermal analyses of minichannels and use of mathematical and numerical models. *Numerical Heat Transfer, Part A: Applications*, *77*(5), 497–537.
11. Z. Said, A. Allagui, M. A. Abdelkareem, H. Alawadhi, & K. Elsaid (2018). Acid-functionalized carbon nanofibers for high stability, thermoelectrical and electrochemical properties of nanofluids. *Journal of Colloid and Interface Science*, *520*, 50–57.
12. Z. Said, M. H. Sajid, R. Saidur, G. A. Mahdiraji, & N. A. Rahim (2015). Evaluating the optical properties of TiO_2 nanofluid for a direct absorption solar collector. *Numerical Heat Transfer, Part A: Applications*, *67*(9), 1010–1027.
13. Z. Said (2016). Thermophysical and optical properties of SWCNTs nanofluids. *International Communications in Heat and Mass Transfer*, *78*, 207–213.
14. M. H. Ahmadi, A. Baghban, M. Sadeghzadeh, M. Zamen, A. Mosavi, S. Shamshirband, & M. Mohammadi-Khanaposhtani (2020). Evaluation of electrical efficiency of photovoltaic thermal solar collector. *Engineering Applications of Computational Fluid Mechanics*, *14*(1), 545–565.
15. IEA (2019). Renewables 2019, IEA, Paris. https://www.iea.org/reports/renewables-2019
16. P. Raj, & S. Subudhi (2018). A review of studies using nanofluids in flat-plate and direct absorption solar collectors. *Renewable and Sustainable Energy Reviews*, *84*, 54–74.

17. M. S. Khan, M. Abid, & T. A. H. Ratlamwala (2019). Energy, exergy and economic feasibility analyses of a 60 MW conventional steam power plant integrated with parabolic trough solar collectors using nanofluids. *Iranian Journal of Science and Technology, Transactions of Mechanical Engineering, 43*(1), 193–209.
18. M. Abid, M. S. Khan, T. A. Ratlamwala, & K. P. Amber (2020). Thermo-environmental investigation of solar parabolic dish-assisted multi-generation plant using different working fluids. *International Journal of Energy Research, 44*(15), 12376–12394.
19. M. S. Khan, M. Yan, H. M. Ali, K. P. Amber, M. A. Bashir, B. Akbar, & S. Javed (2020). Comparative performance assessment of different absorber tube geometries for parabolic trough solar collector using nanofluid. *Journal of Thermal Analysis and Calorimetry, 142*, 2227–2241.
20. M. S. Khan, M. Abid, H. M. Ali, K. P. Amber, M. A. Bashir, &S. Javed (2019). Comparative performance assessment of solar dish assisted s-CO₂ Brayton cycle using nanofluids. *Applied Thermal Engineering, 148*, 295–306.
21. M. S. Khan, K. P. Amber, H. M. Ali, M. Abid, T. A. Ratlamwala, & S. Javed (2019). Performance analysis of solar assisted multigenerational system using therminol VP1 based nanofluids: A comparative study. *Thermal Science, 24*(2A), 865–878.
22. S. L. Song, J. H. Lee, & S. H. Chang (2014). CHF enhancement of SiC nanofluid in pool boiling experiment. *Experimental Thermal and Fluid Science, 52*, 12–18.
23. M. N. Golubovic, H. M. Hettiarachchi, W. M. Worek, & W. J. Minkowycz (2009). Nanofluids and critical heat flux, experimental and analytical study. *Applied Thermal Engineering, 29*(7), 1281–1288.
24. H. D. Kim, J. Kim, & M. H. Kim (2007). Experimental studies on CHF characteristics of nano-fluids at pool boiling. *International Journal of Multiphase Flow, 33*(7), 691–706.
25. T. Yousefi, F. Veisy, E. Shojaeizadeh, & S. Zinadini (2012). An experimental investigation on the effect of MWCNT-H₂O nanofluid on the efficiency of flat-plate solar collectors. *Experimental Thermal and Fluid Science, 39*, 207–212.
26. N. I., ShamsulAzha, H. Hussin, M. S. Nasif, & T. Hussain (2020). Thermal Performance Enhancement in Flat Plate Solar Collector Solar Water Heater: A Review. *Processes, 8*(7), 756.
27. S. K. Das, S. U. Choi , & H. E. Patel (2006). Heat transfer in nanofluids—a review. *Heat Transfer Engineering.* 27, 3–19.
28. Y Xuan, & Li, Q. (2000). Heat transfer enhancement of nanofluids. *International Journal of Heat and Fluid Flow, 21*, 58–64.
29. S. Ferrouillat, A. Bontemps, O. Poncelet, Soriano, O., & Gruss, J-A (2013). Influence of nanoparticle shape factor on convective heat transfer and energetic performance of water-based SiO2 and ZnO nanofluids. *Applied Thermal Engineering, 51*, 839–851.
30. A. J. Moghadam, M. Farzane-Gord, M. Sajadi, & M. Hoseyn-Zadeh (2014). Effects of CuO/water nanofluid on the efficiency of a flat-plate solar collector. *Experimental Thermal and Fluid Science, 58*, 9–14.
31. Said, Z., Saidur, R., Sabiha, M. A., Hepbasli, A., & Rahim, N. A. (2016). Energy and exergy efficiency of a flat plate solar collector using pH treated Al2O3 nanofluid. *Journal of Cleaner Production, 112*, 3915–3926.
32. Z. Said, R. Saidur, N. A. Rahim, & Alim, M. A. (2014). Analyses of exergy efficiency and pumping power for a conventional flat plate solar collector using SWCNTs based nanofluid. *Energy and Buildings, 78*, 1–9.
33. N. I. Shamsul Azha, H. Hussin, M. S. Nasif, & T. Hussain (2020). Thermal performance enhancement in flat plate solar collector solar water heater: A review. *Processes, 8*(7):756.

34. O. Mahian, L. Kolsi, M. Amani, P. Estellé, G. Ahmadi, C. Kleinstreuer, J. Marshall, R. A. Taylor, E. Abu-Nada, S. Rashidi, H. Niazmand, S. Wongwises, T. Hayat, A. Kasaeian, & L. Pop (2018). Recent advances in modeling and simulation of nanofluid flows—Part II: Applications. *Physics Reports*, *791*, 1–59.

35. S. Das, S. Choi, W. Yu, & T. Pradeep (2007). *Nanofluids: Science and Technology*: John Wiley & Sons.

36. A. H. Al-Waeli, K. Sopian, M. T. Chaichan, H. A. Kazem , H. A Hasan, & A. N. Al-Shamani (2017). An experimental investigation of SiC nanofluid as a base-fluid for a photovoltaic thermal PV/T system. *Energy Conversion and Management*, *147*, 547–558.

37. A. H. Al-Waeli, M. T. Chaichan, Kazem, H. A. , & Sopian, K. (2017). Comparative study to use nano-(Al2O3, CuO, and SiC) with water to enhance photovoltaic thermal PV/T collectors. *Energy Conversion and Management*, *148*, 963–973.

38. M. Sardarabadi, M. Passandideh-Fard, & S. Z. Heris (2014). Experimental investigation of the effects of silica/water nanofluid on PV/T (photovoltaic thermal units). *Energy*, *66*, 264–272.

39. M. Sardarabadi, M. Hosseinzadeh, A. Kazemian, & M. Passandideh-Fard (2017). Experimental investigation of the effects of using metal-oxides/water nanofluids on a photovoltaic thermal system (PVT) from energy and exergy viewpoints. *Energy* 138, 682–695.

40. A. Farzanehnia, & M. Sardarabadi (2019). Exergy in photovoltaic/thermal nanofluid-based collector systems. *Exergy and Its Application-Toward Green Energy Production and Sustainable Environment*.

41. Malika, M., Jhadav, P. G., Parate, V. R., & Sonawane, S. S. (2022). Synthesis of magnetite nanoparticle from potato peel extract: its nanofluid applications and life cycle analysis. *Chemical Papers*, 1–14.

42. Malika, M., & Sonawane, S. S. (2022). MSG extraction using silicon carbide-based emulsion nanofluid membrane: Desirability and RSM optimisation. *Colloids and Surfaces A: Physicochemical and Engineering Aspects*, *651*, 129594.

43. Potdar, S. B., Saudagar, P., Potoroko, I., Bagale, U., Sonawane, S., & Sonawane, S. H. (2022). Recent Advances and Reports on Encapsulation in the Food Matrix: A Review. *Current Pharmaceutical Biotechnology*.

44. M. A. Sabiha, R. Saidur, S. Hassani, Z. Said, & S. Mekhilef (2015). Energy performance of an evacuated tube solar collector using single walled carbon nanotubes nanofluids. *Energy Conversion and Management*, *105*, 1377–1388.

45. S. S. Park, & N. J. Kim (2014). A study on the characteristics of carbon nanofluid for heat transfer enhancement of heat pipe. *Renewable energy*, *65*, 123–129.

46. M. Beer, R. Rybár, M. Cehlár, S. Zhironkin, & P. Sivák (2020). Design and numerical study of the novel manifold header for the evacuated tube solar collector. *Energies*, *13*(10), 2450.

47. Gujar, J., Patil, S., & Sonawane, S. (2024). Review on the Encapsulation, Microencapsulation, and Nano-Encapsulation: Synthesis and Applications in the Process Industry for Corrosion Inhibition. *Current Nanoscience*, *20*(3), 314–327.

48. Malika, M., & Sonawane, S. (2024). A Review on the Application of Nanofluids in Enhanced Oil Recovery. *Current Nanoscience*, *20*(3), 328–338.

49. H. Tyagi, P. Phelan, & R. Prasher (2009). Predicted efficiency of a low-temperature nanofluid-based direct absorption solar collector.

50. Gujar, J. G., Patil, S. S., & Sonawane, S. S. (2023). A Review on Nanofluids: Synthesis, Stability, and Uses in the Manufacturing Industry. *Current Nanomaterials*, *8*(4), 303–318.

51. Malika, M., Pargaonkar, A., & Sonawane, S. S. (2023). Application of emulsion nanofluid membrane for the removal of methylene blue dye: stability study. *Chemical Papers*, 1–11.

52. Thakur, P. P., Sonawane, S. S., & Mohammed, H. A. (2023). Recent Trends in Applications of Nanofluids for Effective Utilization of Solar Energy. *Current Nanoscience*, *19*(2), 170–185.
53. Hakke, V. S., Gaikwad, R. W., Warade, A. R., Sonawane, S. H., Boczkaj, G., Sonawane, S. S., & Sapkal, V. S. (2023). Artificial neural network prophecy of ion exchange process for Cu (II) eradication from acid mine drainage. *International Journal of Environmental Science and Technology*, 1–12.
54. Malika, M., Jhadav, P. G., Parate, V. R., & Sonawane, S. S. (2023). Synthesis of magnetite nanoparticle from potato peel extract: its nanofluid applications and life cycle analysis. *Chemical Papers*, *77*(2), 1081–1094.
55. Choudhary, M., Singh, D., Jain, S. K., Sonawane, S. R. S., Singh, D., Devnani, G. L., & Srivastava, K. (2023). Kinetics modeling & comparative examine on thermal degradation of alkali treated Crotalaria juncea fiber using model fitting method. *Journal of the Indian Chemical Society*, 100918.
56. Kale, P., Pujari, S., Gujar, J. G., Sontakke, R., Haddadi, E. I., & Sonawane, S. S. (2023). Batch distillation for separating the acetone and n-heptane binary azeotrope mixture: Optimization and simulation. *Journal of the Indian Chemical Society*, *100*(1), 100795.
57. Kale, P., Pujari, S., Gujar, J. G., Sontakke, R., Haddadi, E. I., & Sonawane, S. (2022). Batch distillation for separating the acetone and n-heptane binary azeotrope mixture: Optimization and simulation. *Journal of the Indian Chemical Society*, 100795.
58. A. Z. Sahin, M. A. Uddin, B. S. Yilbas, & A. Al-Sharaf(2019). Performance enhancement of solar energy systems using nanofluids: An updated review. 10.1016/j.renene.2019.06.108
59. A. K. Tiwari, P. Ghosh, & J. Sarkar (2013). Solar water heating using nanofluids – A comprehensive overview and environmental impact analysis. ISSN 2250-2459 ISO 9001:2008.
60. S. Amalraj, & P. A. Michael (2019). Synthesis and characterization of Al_2O_3 and CuO nanoparticles into nanofluids for solar panel applications. 10.1016/j.rinp.2019.102797
62. M. U. Sajid, & Y. Bicer (2020). Nanofluids as solar spectrum splitters: A critical review. 10.1016/j.solener.2020.07.009
61. K. S. Chaudhari, & P. V. Walke (2014). Applications of nanofluid in solar energy – A review. *International Journal of Engineering Research*, *3*(3), 460–463.
63. R. Kirilov, C. Girginov, & P. Stefchev (2013). black nanofluids for solar absorption on the basis of hydrogen peroxide treated carbon particles. 10.12732/ijpam.v2i2.1
64. W. Chen, C. Zou, X. Li, & L. Li (2016). Experimental investigation of SiC nanofluids for solar distillation system: Stability, optical properties and thermal conductivity with saline waterbased fluid. 10.1016/j.ijheatmasstransfer.2016.11.048
65. H. Wanga, W. Yanga, b, L. Chenga, C. Guana, & H. Yan (2017). Chinese ink: High performance nanofluids for solar energy. 10.1016/j.solmat.2017.10.023
66. M. Yang, S. Wang, Y. Z, R. A. Taylor, M. Moghimi & Y. Wang (2020). Thermal stability and performance testing of oil-based CuO nanofluids for solar thermal applications. 10.3390/en13040876
67. O. Mahian, A. Kianifar, A. Z. Sahin, & S. Wongwises (2014). Performance analysis of a minichannel-based solar collector using different nanofluids. 10.1016/j.enconman.2014.08.021
68. K. Afzal, & A. Aziz (2016). Transport and heat transfer of time dependent MHD slip flow of nanofluids in solar collectors with variable thermal conductivity and thermal radiation. 10.1016/j.rinp.2016.09.017

69. P. Cao, Y. Li, Y. Wu, H. Chen, J. Zhang, L. Cheng, & T. Niu (2019). Role of base fluid on enhancement absorption properties of Fe_3O_4/ionic liquid nanofluids for direct absorption solar collector. 10.1016/j.solener.2019.11.039

70. M. Chen, Y. He, J. Huang, & J. Zhu (2016). Investigation into Au nanofluids for solar photothermal conversion. 10.1016/j.ijheatmasstransfer.2017.01.005

71. W. Chen, C. Zou, X. Li, & H. Liang (2017). Application of recoverable carbon nanotube nanofluids in solar desalination system: An experimental investigation. 10.1016/j.desal.2017.09.025

72. G. Colangelo, E. Favale, P. Miglietta, M. Milanese, & A. de Risi (2015). Thermal conductivity, viscosity and stability of Al_2O_3-diathermic oil nanofluids for solar energy systems. 10.1016/j.energy.2015.11.032

73. E. Ebrahimnia-Bajestan, M. Charjouei Moghadam, H. Niazmand, W. Daungthongsuk, & S. Wongwises (2015). Experimental and numerical investigation of nanofluids heat transfer characteristics for application in solar heat exchangers. 10.1016/j.ijheatmasstransfer.2015.08.107

74. M. Faegh, & M. B. Shafi (2017). Experimental investigation of a solar still equipped with an external heat storage system using phase change materials and heat pipes. 10.1016/j.desal.2017.01.023

75. Q. He, S. Wang, S. Zeng, & Z. Zheng (2013). Experimental investigation on photothermal properties of nanofluids for direct absorption solar thermal energy systems. 10.1016/j.enconman.2013.04.019

76. Y. Hu, Y. He, H. Gao, & Z. Zhang (2019). Forced convective heat transfer characteristics of solar salt-based SiO_2 nanofluids in solar energy applications. 10.1016/j.applthermaleng.2019.04.109

77. S. Ladjevardi, A. Asnaghi, P. Izadkhast, & A. Kashani (2012). Applicability of graphite nanofluids in direct solar energy. 10.1016/j.solener.2013.05.012

78. S. Manikandan, & K. Rajan (2015). Sand-propylene glycol-water nanofluids for improved solar energy collection. 10.1016/j.energy.2016.07.120

79. S. Mesgari, S. Coulombe, N. Hordy, & R. A. Taylor (2016). Thermal stability of carbon nanotube-based nanofluids for solar thermal collectors. 10.1179/1432891714Z.0000000001169

80. D. Rativa, & L. A. Gómez-Malagón (2015). Solar radiation absorption of nanofluids containing metallic nanoellipsoids. 10.1016/j.solener.2015.05.048

81. E. Sani, N. Papi, L. Mercatelli, & G. Żyła (2018). Graphite/diamond ethylene glycol-nanofluids for solar energy applications. 10.1016/j.renene.2018.03.078

82. N. A. Sheikh, F. Ali, I. Khan, M. Gohar, & M. Saqib (2018). On the applications of nanofluids to enhance the performance of solar collectors: A comparative analysis of Atangana-Baleanu and Caputo-Fabrizio fractional models. 10.1140/epjp/i2017-11809-9

83. R. A. Taylor, P. E. Phelan, T. P. Otanicar, C. A. Walker, M. Nguyen, S. Trimble, & R. Prasher (2013). Applicability of nanofluids in high flux solar collectors. 10.1063/1.3571565

84. J. P. Vallejo, L. Mercatelli, M. R. Martina, A. Di Rosa, A. Dell'Oro, L. Lugo, & E. Sani (2019). Comparative study of different functionalized graphene-nanoplatelet aqueous nanofluids for solar energy applications. 10.1016/j.renene.2019.04.075

85. S. K. Verma, A. K. Tiwari, & D. S. Chauhan (2016). Performance augmentation in flat plate solar collector using MgO/water nanofluids. 10.1016/j.enconman.2016.07.007

86. N. Chen, H. Ma, Y. Li, J. Cheng, C. Zhang, D. Wu, & H. Zhu (2016). Complementary optical absorption and enhanced solar thermal conversion of CuO-ATO nanofluids. 10.1016/j.solmat.2016.12.049

87. M. Turkyilmazoglu (2016). Performance of direct absorption solar collector with nanofluid mixture. *Energy Conversion and Management, 114*, 1–10.

3 Application of Nanofluids for the Car Radiator

3.1 INTRODUCTION

Efficient car radiators are important for the better performance of the engine. Thus, for this purpose, the Application of nanofluids in the thermal management of car radiators is explored by various researchers. Saidur et al. [1] reviewed the application of nanofluids in various sectors and drew attention to the possible application of nanofluids in the automobile industry. As we have discussed in Chapter 1, the concentration of nanoparticles and suitable base fluids are important for efficient nanofluids. The author has reported the various advantages of the nanofluids. Peyghambarzadeh et al. [2] performed various experiments using the alumina nanofluids. Ethylene glycol is added to the water as an anti-knocking agent. Experiments are conducted at various flow rates and different nanoparticle concentrations. Nearly 40% enhancement is reported for this nanofluids system. Figure 3.1 represents the variation of the Nusselt number for different Reynolds numbers.

Ahmed et al. [3] conducted experiments with TiO_2/water nanofluids for the car radiators using various nanoparticle concentrations. The flow rate of nanofluids is maintained in the laminar region. Authors have reported that the 0.2 vol% nanoparticle concentration has increased the car radiator performance by 47%. TiO_2 nanoparticle is considered a good nanoparticle to use in nanofluids. This is because of the lower specific gravity and good thermal conductivity. Hafiz Muhammad Ali et al. [4] reported a 46% enhancement in heat transfer by ZnO-based nanofluids. The optimum nanoparticle concentration that, the authors have reported is 0.2 vol%. Nanofluids with nanoparticle concentration above this concentration have comparatively less heat transfer. The inlet temperature is maintained constant for all the studies. Naraki et al. [5] used CuO-based nanofluids and the optimum nanoparticle concentration recorded for these nanofluids is 0.4 vol%. But the heat transfer was only 8%. Hussein et al. [6] used TiO_2 and SiO_2-based nanofluids for the car radiators. The maximum heat transfer enhancement is recorded with the silica-based nanofluids than TiO_2-based nanofluids. 11% heat transfer enhancement is recorded with the TiO_2-based nanofluids, while a 22% increase in the heat transfer is recorded with the silica-based nanofluids [7].

Hashemabadi et al. [8] conducted the experiments with the alumina/water nanofluids. A 45% increase in heat transfer is recorded with this nanofluid system. Figure 3.2 represents the nanofluids performance for increasing the Reynolds number. Saidur et al. [9] used Cu nanoparticles for the heat transfer enhancement

DOI: 10.1201/9781003404767-3

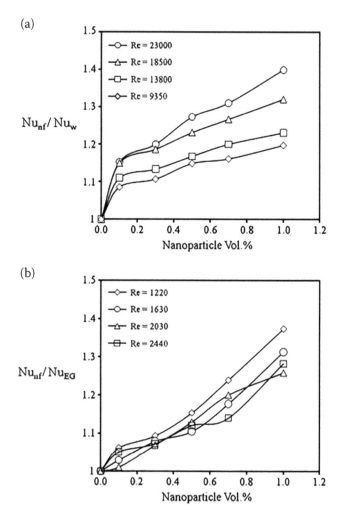

FIGURE 3.1 Variation of Nusselt number for different Reynolds numbers (a) water-based nanofluids (b) ethylene glycol-based nanofluids [2].

but only a 3.8% increase in the heat transfer is recorded. Zou et al. [10] conducted experiments with the SiC-based nanofluids and an increase in the thermal conductivity by 54% was recorded at the 0.5% nanoparticles concentration. Subhedar et al. [11] conducted experiments with alumina nanoparticles in the water and ethylene glycol mixture. A 30% increase in the heat transfer rate was reported Oliveira et al. [12] conducted experiments with MWCNT-based nanofluids and reported a decrease in the heat transfer by 17%. Thus the addition of nanoparticles doesn't always increase the heat transfer rate. Goudarzi et al. [13] conducted an experiment with alumina/ethylene glycol with wire coil inserts in the radiators. The heat transfer rate is increased by 9% than ethylene glycol alone as a base fluid. Tijani et al. [14] compared the performance of alumina-based nanofluids and CuO-based nanofluids by experiments and mathematical models. Results

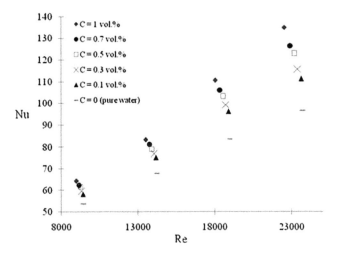

FIGURE 3.2 Variation of Nusselt number with increasing Reynolds number [8].

obtained for CuO-based nanofluids were better than the alumina-based nanofluids. Figure 3.3 represents the various results obtained from the alumina-based nanofluids and CuO-based nanofluids. Figure 3.4 is a representation of numeric results obtained from CATIA and other simulation tools.

Filho et al. [15] conducted experiments with the silver and graphene nanoparticles and reported that the silver nanoparticle-based nanofluids reported an enhancement of 4% while graphene-based nanofluids reported a decrease in the heat transfer performance. Elbadawy et al. [16] compared the performance of alumina-based nanofluids and CuO-based nanofluids using the numerical approaches. Ali et al. [17] performed the experimentations with the ZnO/water

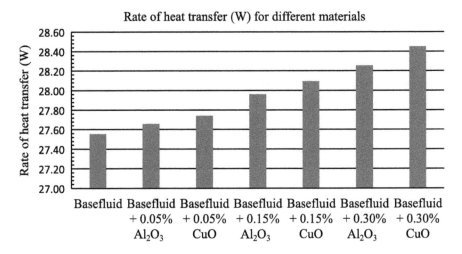

FIGURE 3.3 Rate of heat transfer for different concentrations of nanofluids of alumina and CuO [14].

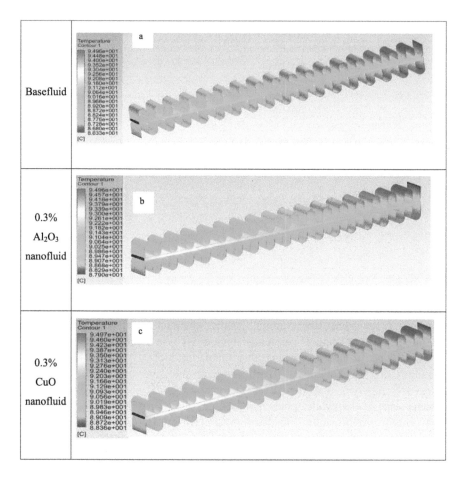

FIGURE 3.4 Different results obtained for different working fluids [14].

nanofluids for the car radiator. An increase in the heat transfer by 46% is achieved. Peyghambarzadeh et al. [18] also used alumina and CuO-based nanofluids to evaluate the performance of the ε-NTU method. A 9% increase in heat transfer is reported by this method. Ali et al. [19] reviewed the various literature regarding this topic and reported the effect of various parameters like size, the morphology of particles, nanoparticle loading, nanofluids velocity, etc. on heat transfer. Reddy et al. [20] conducted the experimentations by TiO$_2$/water-based nanofluids and reported a 37% increase in the performance compared to the base fluid. Shashishekar et al. [21] conducted experiments with functionalized MWCNT and water/ethylene glycol. A 45% increase in the heat transfer is reported. Yang et al. [22] reviewed various literature on the modeling of car radiators using artificial neural networks. Sidik et al. [23] reviewed various literature regarding the recent trends of nanofluids application in car radiators. Harish et al. [24] conducted experiments with graphene-based nanofluids for car radiator applications. The heat transfer enhancement is from 20 to 51% for various operating conditions. Chiavazzo et al. [25] and

Sidik et al. [26] also reviewed the numerical approaches developed for the nanofluids application in the car radiator. Sidik et al. [27] conducted experiments with Perodua Kelisa 1000 cc radiators using MWCNT-based nanoparticles. The enhancement reported is 196.3%. The radiator geometry is represented in Figure 3.5. Hussein et al. [28] used silica-based nanofluids and a 50% increase was reported.

Various industries are directly dependent on the automobile sector. The development of the automobile sector is considered one of the indicators of a country's overall growth. Scientists are developing efficient cooling systems for the engines used in automobiles [29]. The development of efficient car radiators is essential for the sustainable utilization of resources because a bad cooling system

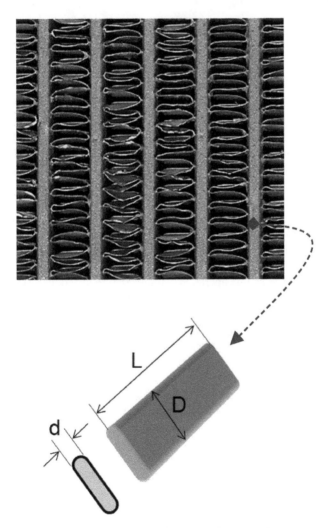

$$d = 1.8 \ mm \ ; D = 15.5 \ mm \ ; L = 137 \ mm$$

FIGURE 3.5   The fin and tube of car radiator geometry [27].

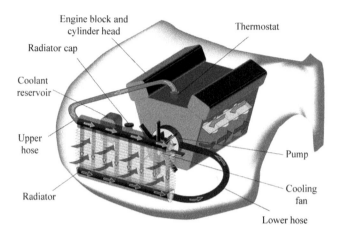

FIGURE 3.6 Automobile thermal management systems [39].

leads to a decrease in the overall performance of automobiles [30]. The heat generated in the engine operation is absorbed by the vehicle's cooling system. Thermal cooling is also important to maintain the engine surface in good condition. Recent studies have shown various strategies to improve the engine performance [31]. It is also important to maintain the good health of the engine to reduce the environmental impact of the automobile industry [32]. Nanofluids are one such technology to improve the thermal management system of automobiles [33].

The schematic representation of the thermal management system is shown in Figure 3.6. Cooling fan, hose for coolant, radiator and thermostat are important components of this system. Coolant is passed around the engine surface and this coolant is traveled within the thermal management system by hose. The heat generated during the engine operation is transferred to the coolant. Then this coolant passes from the upper hose of the radiator and comes to the lower hose of the radiator. Radiator decreases the temperature of coolant by convection heat transfer. The heat is transferred to the surrounding air from the coolant. The coolant is again sent to the engine cooling chamber by pump [34]. The radiator is an important part of this system. Radiators are nothing but a special type of heat exchanger [35,36]. The radiators are made of tubes and fins within it. Fins and tubes absorb heat from coolant by conduction and heat is transferred channel and fins to the surroundings by forced convection [37]. Brass, copper, and aluminium are generally used to manufacture the radiators. Among all these, aluminium is generally used as a material of construction for radiators [38].

The performance of automobile thermal cooling systems is dependent on various factors such as radiator design, fouling within the channels, and operating conditions. The component for the heat transfer is air-side heat transfer resistance. Coolant change will not change the coolant temperature. Thus, it is also important to study the effect of other parameters like air and coolant mass flow rate, inlet air temperature and pitch of fins and tube. In some cases, it is observed that, due to an increase in the height of the radiator, air density decreases and air mass flow rate gets affected [40]. Usually, the performance of the car radiators decreases in the summer. The air temperature is higher during the summer. The performance of the thermal cooling system also needs to withstand conditions when the engine is producing maximum power. For example, when the vehicle continuously working for hours or days. The thermal cooling system may fail when the engine is generating maximum power. For example, when the automobile is climbing the hill. While designing the cooling system we must consider all these parameters.

Coolants for automobiles must have less freezing point. Because the vehicle has to work in low-temperature regions like the Himalayas. Coolants also must have a high boiling point to work in deserts where the temperature is too high. Thus ethylene glycol is added as an additive for this purpose. Though water has more thermal conductivity than ethylene glycol. Ethylene glycol has a higher boiling point than water and a lower freezing point. Thus, a suitable mixture of water and ethylene glycol is used in the thermal management system. Composition mainly depends on the working conditions of the automobile. For example, Gollin and Bjork [41,42] conducted experiments with pure water, pure propylene glycol, a mixture of propylene glycol and water and a mixture of ethylene glycol and water at different compositions. For better data, experiments are conducted at the five different radiators with different areas from 350 to 530 square feet. All these experiments are conducted for four different flow rates. The temperature difference of 60 °C is maintained throughout the experiments between nanofluids inlet temperature and airflow temperature. From experiments, the author proved that water is the best coolant for automobile engine cooling.

3.2   RECENT ADVANCES IN EXPERIMENTAL STUDIES

In the last few decades, researchers have used various nanofluid systems for the cooling of engines in automobiles. As we have discussed in the previous section, nanofluids offer better heat transfer than conventional coolants. The lab-scale experimental setup for the study of nanofluids application in the car radiator is shown in Figure 3.7. Rotameters are used to measure the nanofluids flow rate. A fan speed controller is used

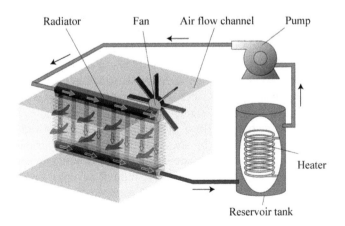

FIGURE 3.7   Lab-scale experimental setup of nanofluids application in car radiator [39].

to manipulate or regulate the airspeed. J-type thermocouples are used for the measurement of temperature on radiator walls. The heating tank is used to heat the nanofluids. Generally, heating is done to maintain the constant temperature.

Lockwood et al. [43] from Valvoline Company conducted experimentations with the graphite/engine oil nanofluids for the thermal cooling system of IC engines and reported that the nanofluids with better thermal conductivity should be preferred for the car radiator application. Saripella et al. [44] conducted the experimentations with the CuO nanoparticles in a mixture of water and ethylene glycol base fluid. They used Volvo truck engines for these experiments and thus, it highlighted that the nanofluids type is important for the efficient working of the car radiator. The study also showed that these efficient nanofluids are important for the decrease in the fuel consumption of the engine for the same operation. Researchers from the Delphi Company like, Wallner et al. [45] conducted the research with nanofluids for the increase in the performance of IC engines with a reduction in the size of the engine. Vajjha et al. [46] conducted experiments with alumina and CuO-based nanofluids for various car radiator applications and reported the enhancement in the heat transfer with an increase in nanoparticle loading. Similar results are reported by other researchers like Leong et al. [47]. They have reported an increase in heat transfer coefficient with an increase in nanoparticle loading in the nanofluids. Peyghambarzadeh et al. [48] also used alumina-based nanofluids and reported the heat transfer efficiency increased by 45%. Hussein et al. [49], reported that the nanoparticles used in nanofluids increase the friction factor. This friction factor increases with the increase in nanoparticle concentration.

Chougule et al. [50] compared the heat transfer performance of MWCNT-based nanofluids and alumina-based nanofluids. At the same concentration of nanoparticles at 1 vol%. The performance of MWCNT-based nanofluids is increased by 91% and the performance of alumina-based nanofluids is increased by 52%. Leong et al. [47] used CuO nanoparticles and ethylene glycol as a base fluid for the investigation of nanofluids application in car radiators. Various data required for the calculation are taken from the literature. The increase in the heat transfer performance. But, pumping power is increased by 12%. However, the area exposed to air can be reduced to 18.7% to achieve the same heat transfer. Researchers have used various mathematical strategies to evaluate the performance of nanofluids in car radiators by calculating the overall heat transfer coefficient. One such attempt was made by Peyghambarzadeh et al. [51]. They used the conventional NTU technique. Authors reported that the stability of nanofluids is a very important parameter to ensure better heat transfer. Authors reported that, at a 10.1 pH value, CuO-based nanofluids are more stable and at an 11.1 pH value, Fe_2O_3-based nanofluids are more stable. The use of hybrid nanofluids for the car radiator application is also investigated by the researchers. Hybrid nanofluids can be synthesized by adding two or more nanoparticles or by doping one nanoparticle on another nanoparticle and then mixing it in base fluid. Ramalingam et al. [52] doped alumina nanoparticles on milled silicon carbide nanoparticles and un-milled silicon carbide nanoparticles. The base fluid of water and ethylene glycol with equal composition is used. These nanoparticles and base fluids are used in IC engines as a coolant. The author reported that the thermal conductivity of hybrid nanofluids of Milled silicon carbide is better than hybrid nanofluids of unmilled silicon carbide. As the random movement of nanoparticles increases, the nanoparticle concentration. The Nusselt number increases by 23% for 0.8 vol% concentration of alumina-doped silicon carbide-based hybrid nanofluids. Koçak et al. [53] synthesized various hybrid nanofluids of different compositions and concentrations of silver and copper nanoparticles doped on the TiO_2 nanoparticle. The base fluid used is a mixture of water and ethylene glycol of equal composition. The experimentations with these hybrid nanofluids are conducted in the louver corrugated car radiators. When 0.3% silver doped TiO_2 nanoparticle is used in 2% concentration, thermal conductivity increases by 24% and these nanofluids result in a 3% increase in pumping power. These hybrid nanofluids also reported an increase of 28% in heat transfer coefficient. Figure 3.8 represents the geometrical details used for the experimental setup and the real-time setup used for the study is shown in Figure 3.9.

Kumar et al. [54] used a wavy finned radiator for the experimental study of hybrid nanofluids. Researchers used nanoparticles of alumina, graphene and carbon nanotubes of different morphology. Hybrid nanofluids of alumina and graphene have shown the best result in terms of thermal conductivity enhancement. Sahoo et al. [55] conducted the experimental analysis of hybrid nanofluids of alumina and one nanoparticle from five nanoparticles (i.e., silver, copper, silicon carbide, copper

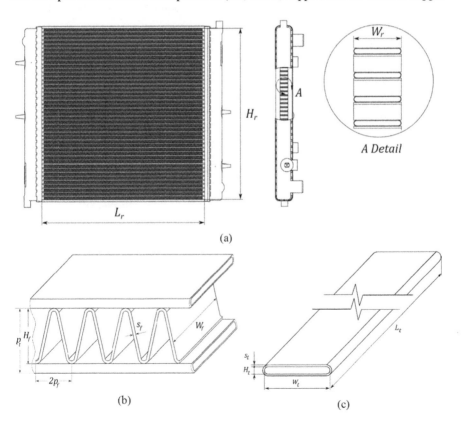

FIGURE 3.8 Geometrical details of the car radiator for experimentation [53].

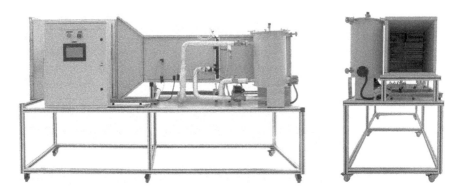

FIGURE 3.9   Real-time images of the car radiator-based cooling system [53].

oxide and nanoparticles). Silver/alumina nanofluids at 1 vol% nanoparticle showed the best performance in terms of heat transfer rate. Hybrid nanofluids of alumina and silicon carbide have shown the best results in terms of performance index. Li et al. [56] conducted the experimentations in a car radiator with silicon carbide nanoparticles doped by the multi-walled carbon nanotubes in ethylene glycol. At 0.4 vol% of this hybrid nanoparticle, the observed thermal conductivity was maximum. The

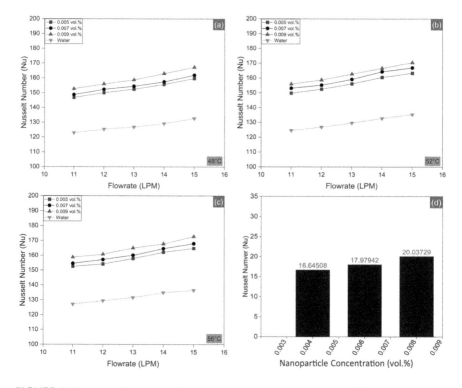

FIGURE 3.10   (a-c) Variation of Nusselt number for flow rate at different temperature values (d) Increment of Nusselt number compared to the water [57].

TABLE 3.1

Experimental Findings for Car Radiator Using Nanofluids [39]

Sr. No.	Nanofluid Systems	Findings	References
1	CuO, Al$_2$O$_3$, anti-foam-transmission oil	CuO-based nanofluid gives better results at various speeds.	58
2	CuO/coolant	The cooling capacity of nanofluid with a 3% volume concentration increases by 15%.	59
3	CuO/Ethylene glycol+ water (50:50)	At lower temperatures, the performance of the engine improves.	60
4	Al$_2$O$_3$/water	At 1 volume % of nanoparticles concentration; heat performance is best.	61
5	Al$_2$O$_3$/Ethylene glycol	Maximum thermal conductivity improvement of 4.5% is observed at 1.5 vol% at 50 °C.	62
6	Al$_2$O$_3$/Ethylene glycol	Maximum thermal conductivity improvement of 11.25% is observed at 3.5 vol% at 80 °C.	63
7	Al$_2$O$_3$/Ethylene glycol +water (50:50)	Maximum thermal conductivity improvement of 8.3% is observed at 1 vol% at 50 °C.	64

increase in the thermal conductivity was 32% higher than the thermal conductivity of the base fluid. An increase in the temperature leads to a decrease in the performance and also the excessive nanoparticle concentration. The heat transfer coefficient increases by 28% for this nanofluid system. Abbas et al. [57] developed a hybrid nanofluid made up of Fe$_2$O$_3$ and TiO$_2$ with equal composition and water is used as base fluid. The concentration of nanofluids is varied for each experiment. The temperature of nanofluids is increased from 46 °C to 56 °C, it is observed that an increase in the nanofluids temperature leads to an increase in the heat transfer efficiency. An 8% increase in the heat transfer is observed. These results are shown in Figure 3.10. Studies reported by other researchers are tabulated in Table 3.1.

3.3 CASE STUDY: APPLICATION OF COBALT OXIDE NANOFLUIDS IN THE CAR RADIATOR

3.3.1 INTRODUCTION

As we have discussed in Chapter 1, Nanofluids can be synthesized in two ways, the first is a step method and the second is a two-step method. Cobalt oxide is synthesized by the two-step method. i.e., cobalt oxide nanoparticles are synthesized first then these nanoparticles are added to the base fluid of water and ethylene glycol. The nanoparticles are synthesized by the method available in the literature. These nanoparticles are added to the base fluid in the required amount.

Figure 3.11 is the schematic representation of lab scale experimentation setup for the nanofluids application for car radiator. These experiments were conducted in the nanotechnology laboratory, VNIT, Nagpur. The numeric validation of these results was also conducted. The basic geometry data required for these simulations are tabulated in Table 3.2. The car radiator used for the study is connected to the hot

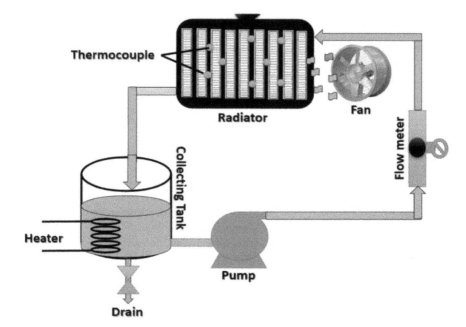

FIGURE 3.11 Lab scale setup of the nanofluids application in car radiator.

TABLE 3.2
Geometrical Configurations of the Car Radiator Model

Tube length	29 cm
Tube thickness	0.5 cm
Tube height	0.14 cm
Tube width	0.2 cm
Number of fins	125
Space between tubes	1.5 cm
Tube hydraulic diameter	0.261 cm
Material used	aluminium
Cell count	1,647,980

water source. The hot water is allowed to pass through the tubes of the radiator. This hot water is maintained at a constant temperature for the given set of experiments. The temperature at the hot water source and end of the radiator is measured by the thermocouples. A set of thermocouples are used to measure the wall temperature of the radiator. The heat transfer is mainly due to the convection mode of heat transfer from the wall of the radiator to the surrounding air. Thus, the ambient temperature and wall temperature are important. The flow rate of air is also maintained constant for the set of experimentations.

3.4   DEVELOPMENT OF SIMULATION MODEL

Generally, the radiator geometries are made in three stages: firstly a rectangular tube is made as shown in Figure 3.12(a). Then, the fins are attached to this tube as shown in Figure 3.12(b). And the mesh is applied for the computation as shown in Figure 3.12(c). As we know, the grid independence test should be conducted before using any grid size. In the case of the current case study, the software used for these simulations is ANSYS ICEM-CFD software.

3.4.1   DEVELOPMENT OF MATHEMATICAL MODEL AND CALCULATION OF THE PHYSICAL PROPERTIES

Various equations available in the literature are used for the calculation of nanofluids properties in the car radiator. We have already seen the geometrical properties used for the current case study. Similarly, we require thermo-physical properties of the base fluid and nanoparticles to calculate the thermo-physical properties of nanofluids. For this purpose, the values available for the nanoparticles and base fluid in the literature are used. These values are tabulated in Table 3.3.

Various thermo-physical properties of nanofluids are calculated by the following equations.

The value of viscosity of nanofluids is calculated by the following equation [65]

$$\mu_{nf} = (1 + 2.5\varphi)\mu_{bf} \qquad (3.1)$$

where μ_{nf} means the viscosity of nanofluids, μ_{bf} stands for the viscosity of base fluids; φ: Particle volume fraction

The value of density of nanofluids is calculated by following the equation [66]

$$\rho_{nf} = \varphi\rho_p + (1 - \varphi)\rho_{bf} \qquad (3.2)$$

where ρ_{nf} stands for a density of nanofluid, ρ_p means the density of the particle, ρ_{bf} is the density of the base fluid, and φ represents the volume fraction of the particle [67]

The value of the specific heat of the nanofluids is calculated by the following equation.

$$(\rho C_p)_{nf} = \varnothing(C_p\rho)_p + (1 - \varnothing)(C_p\rho)_{bf} \qquad (3.3)$$

The value of thermal conductivity of nanofluids is calculated by following the equation

$$K_{nf} = \frac{K_p + 2K_{bf} + \varnothing(K_p - K_{bf})}{K_p + 2K_{bf} - \varnothing(K_p - K_{bf})}K_{bf} \qquad (3.4)$$

K_{nf} represents the thermal conductivity of nanofluid, K_p stands for thermal conductivity of particles, K_{bf} is the thermal conductivity of the base fluid [68]

(a)

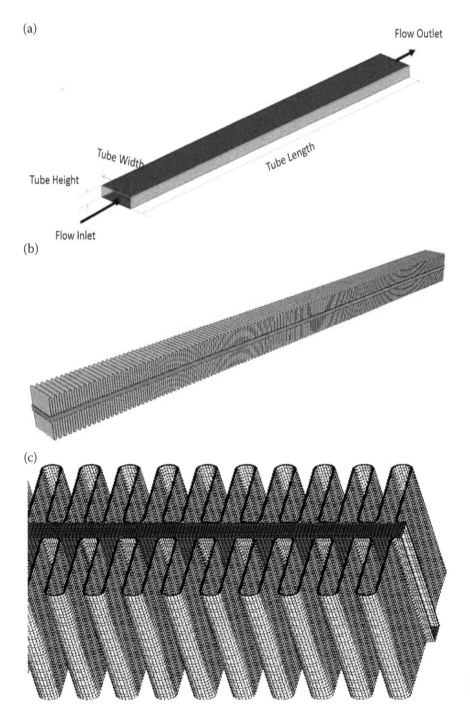

FIGURE 3.12 (a) Schematic of the flat radiator tube. (b) Computational model of the radiator. (c) Details of structured hexahedral mesh.

TABLE 3.3
Various Properties of the Base Fluid and Nanoparticles at 25 °C

Working Fluid	Density (kg/m³)	Specific Heat Capacity (J/kg.K)	Viscosity (Pa.sec)	Thermal Conductivity (W/m.K)
Co_3O_4	6,081	225	–	8.69
Water	995	4,180	0.00089	0.609
Ethylene glycol	1,110	2,200	0.0161	0.256
Base fluid	1,027.01	3,570	0.00076	0.415

The Prandtl number used for the calculation of heat transfer behaviour is calculated by following the equation

$$P_r = \frac{C_p \mu}{k} \qquad (3.5)$$

The Dittus Boelter equation used for the current case study is given as

$$Nu = 0.023 \times (Re)^{0.8} \times (Pr)^{0.3} \qquad (3.6)$$

$$h = \frac{Nu \times k}{d}$$

where h = heat transfer coefficient and k represents thermal conductivity of the fluid.

3.4.2 BOUNDARY CONDITION

The boundary condition for the simulation study needs to be applied at the walls of the radiator, inlet and outlet of the radiator and on the surface of the radiator tube.

- A realizable k-epsilon model of the CFD is used for the current case study of nanofluids application in the car radiator.
- The inlet temperature of the nanofluids is set at nearly 63 °C ± 3 °C. The volumetric flow rate of nanofluids is varied from 2 lpm to 4 lpm at intervalsof 0.5 lpm.
- The boundary condition of the outlet is set as the pressure at atmospheric pressure.
- The boundary conditions of the walls of the tube and fin are varied as per the requirement of flow rate. The ambient temperature is set as 5 °C.
- The nanofluids flow is considered a turbulent flow the calculated thermophysical values are considered as constant. The pressure of the system is considered as a constant.

Temperature (K)

FIGURE 3.13 Variation of the temperature for the length of car radiator geometry pure water at a flow rate of 3 lpm.

- The nanofluids flow and temperature drop are considered constant throughout the geometry. To avoid a large simulation time; only a single tube with the adjacent fins is taken for the simulation study. The heat loss during the experimentation is assumed to be negligible.

3.4.3 MESH INDEPENDENCE TEST

A mesh independence test is conducted to find the optimum mesh size of the geometry. This mesh size allows us to calculate the parameters in the minimum possible time. This study helps to reduce the number of elements of the mesh. In this study, five simulations with different grid sizes are conducted to obtain the coolant output temperature that is independent of the number of elements.

In this study, the number of mesh elements was varied by changing the grid spacing during the mesh generation process. The constant coolant output temperature obtained was 316.14 K. A lower number of mesh elements requires lesser computing time as compared to that of a higher mesh element. Thus, a mesh having 1,647,980 elements (the smallest among all) was chosen in the CFD simulation. Figure 3.13 represents the CFD model and results obtained from the study.

3.5 RESULTS AND DISCUSSION

3.5.1 VALIDATION OF MATHEMATICAL MODELING WITH EXPERIMENTAL DATA

The results obtained with the simulations are compared with the experimental values. This comparison is tabulated in Table 3.4. The table represents the Nusselt number values of the simulation and experimental study. The percent error is also

TABLE 3.4
Comparison of Mathematical and Simulation Results

Flow Rate (lpm)	Experimental				Simulation				Percentage Error			
	Nanoparticles Concentration (v/v)				Nanoparticles Concentration (v/v)				Nanoparticles Concentration (v/v)			
	0	0.05	0.2	0.3	0	0.05	0.2	0.3	0	0.05	0.2	0.3
2	106.5	110.1	111.6	130.6	101.2	103.3	103.7	119.2	5.2	6.6	7.56	9.58
2.5	128.6	131.1	134.7	152.4	121.0	123.5	124.0	135	6.3	6.2	8.64	12.5
3	150.4	148.9	151.2	173.4	140.0	142.9	143.5	151.	7.4	4.2	5.36	14.3
3.5	165.0	163.6	177.3	186.0	158.4	161.6	162.3	170.	4.2	1.2	9.24	12.3
4	190.7	187.7	203.1	211.8	176.2	179.8	180.6	185.3	8.2	4.3	12.4	14.3
						Total Average Error					8.0225%	

given in the table. The overall error observed was nearly 8%. We took only one car radiator tube and in the actual car radiator, there are more tubes present. Thus, the values of experimental heat transfer are more than the values calculated from the simulations. Generally 15% error in the experimental and simulation results is acceptable. The maximum error value reported in Table 3.4 is 14.32%. This is also well below the acceptable range. Thus, the obtained results can be used for further modeling or experiments.

3.6 CONCLUSION

In this chapter, the experimental advances in the application of nanofluids for various car radiators are discussed in detail. The experimental results for the case study are compared with the numeric study. The model for numeric study is developed and a detailed procedure for the model development is reported in this chapter. Nanofluids have shown an increase in performance in nearly every study. Only the stability of nanofluids needs to improve for better performance.

REFERENCES

1. R. Saidur, K. Y. Leong, & H. A. Mohammad (2011). A review on applications and challenges of nanofluids. *Renewable and Sustainable Energy Reviews*, *15*(3), 1646–1668.
2. S. M. Peyghambarzadeh, S. H. Hashemabadi, S. M. Hoseini, & M. S. Jamnani (2011). Experimental study of heat transfer enhancement using water/ethylene glycol based nanofluids as a new coolant for car radiators. *International Communications in Heat and Mass Transfer*, *38*(9), 1283–1290.
3. S. A. Ahmed, M. Ozkaymak, A. Sözen, T. Menlik, & A. Fahed (2018). Improving car radiator performance by using TiO_2-water nanofluid. *Engineering Science and Technology, an International Journal*, *21*(5), 996–1005.
4. H. M. Ali, H. Ali, H. Liaquat, H. T. B. Maqsood, & M. A. Nadir (2015). Experimental investigation of convective heat transfer augmentation for car radiator using ZnO–water nanofluids. *Energy*, *84*, 317–324.
5. M. Naraki, S. M. Peyghambarzadeh, S. H. Hashemabadi, & Y. Vermahmoudi (2013). Parametric study of overall heat transfer coefficient of CuO/water nanofluids in a car radiator. *International Journal of Thermal Sciences*, *66*, 82–90.
6. A. M. Hussein, R. A. Bakar, K. Kadirgama, & K. V. Sharma (2014). Heat transfer enhancement using nanofluids in an automotive cooling system. *International Communications in Heat and Mass Transfer*, *53*, 195–202.
7. S. L. Sharifi, et al. (March 2013). Characterization of cobalt oxide Co_3O_4 nanoparticles prepared by various methods: Effect of calcination temperatures on size, dimension and catalytic decomposition of hydrogen peroxide. *International Journal Nanoscience Nanotechnology*, *9*(1), 51–58.
8. S. M. Peyghambarzadeh, S. H. Hashemabadi, M. S. Jamnani, & S. M. Hoseini (2011). Improving the cooling performance of automobile radiator with Al_2O_3/water nanofluid. *Applied Thermal Engineering*, *31*(10), 1833–1838.
9. K. Y. Leong, R. Saidur, S. N. Kazi, & A. H. Mamun (2010). Performance investigation of an automotive car radiator operated with nanofluid-based coolants (nanofluid as a coolant in a radiator). *Applied Thermal Engineering*, *30*(17–18), 2685–2692.

10. X. Li, C. Zou, & A. Qi (2016). Experimental study on the thermo-physical properties of car engine coolant (water/ethylene glycol mixture type) based SiC nanofluids. *International Communications in Heat and Mass Transfer, 77*, 159–164.

11. D. G. Subhedar, B. M. Ramani, & A. Gupta (2018). Experimental investigation of heat transfer potential of Al_2O_3/water-mono ethylene glycol nanofluids as a car radiator coolant. *Case Studies in Thermal Engineering, 11*, 26–34.

12. G. A. Oliveira, E. M. C. Contreras, & E. P. Bandarra Filho (2017). Experimental study on the heat transfer of MWCNT/water nanofluid flowing in a car radiator. *Applied Thermal Engineering, 111*, 1450–1456.

13. K. Goudarzi, & H. Jamali (2017). Heat transfer enhancement of Al_2O_3-EG nanofluid in a car radiator with wire coil inserts. *Applied Thermal Engineering, 118*, 510–517.

14. A. S. Tijani, & A. S. bin Sudirman (2018). Thermos-physical properties and heat transfer characteristics of water/anti-freezing and Al_2O_3/CuO based nanofluid as a coolant for car radiator. *International Journal of Heat and Mass Transfer, 118*, 48–57.

15. E. M. C. Contreras, G. A. Oliveira, & E. P. Bandarra Filho (2019). Experimental analysis of the thermohydraulic performance of graphene and silver nanofluids in automotive cooling systems. *International Journal of Heat and Mass Transfer, 132*, 375–387.

16. M. Elsebay, I. Elbadawy, M. H. Shedid, & M. Fatouh (2016). Numerical resizing study of Al_2O_3 and CuO nanofluids in the flat tubes of a radiator. *Applied Mathematical Modelling, 40*(13–14), 6437–6450.

17. H. M. Ali, H. Ali, H. Liaquat, H. T. B. Maqsood, & M. A. Nadir (2015). Experimental investigation of convective heat transfer augmentation for car radiator using ZnO–water nanofluids. *Energy, 84*, 317–324.

18. S. M. Peyghambarzadeh, S. H. Hashemabadi, M. Naraki, & Y. Vermahmoudi (2013). Experimental study of overall heat transfer coefficient in the application of dilute nanofluids in the car radiator. *Applied Thermal Engineering, 52*(1), 8–16.

19. M. U. Sajid, & H. M. Ali (2019). Recent advances in application of nanofluids in heat transfer devices: A critical review. *Renewable and Sustainable Energy Reviews, 103*, 556–592.

20. S. Devireddy, C. S. R. Mekala, & V. R. Veeredhi (2016). Improving the cooling performance of automobile radiator with ethylene glycol water based TiO_2 nanofluids. *International Communications in Heat and Mass Transfer, 78*, 121–126.

21. R. Jadar, K. S. Shashishekar, & S. R. Manohara (2017). f-MWCNT nanomaterial integrated automobile radiator. *Materials Today: Proceedings, 4*(10), 11028–11033.

22. N. Zhao, S. Li, & J. Yang (2016). A review on nanofluids: Data-driven modeling of thermalphysical properties and the application in automotive radiator. *Renewable and Sustainable Energy Reviews, 66*, 596–616.

23. N. A. C. Sidik, M. N. A. W. M. Yazid, & R. Mamat (2017). Recent advancement of nanofluids in engine cooling system. *Renewable and Sustainable Energy Reviews, 75*, 137–144.

24. C. Selvam, D. M. Lal, & S. Harish (2017). Enhanced heat transfer performance of an automobile radiator with graphene based suspensions. *Applied Thermal Engineering, 123*, 50–60.

25. M. B. Bigdeli, M. Fasano, A. Cardellini, E. Chiavazzo, & P. Asinari (2016). A review on the heat and mass transfer phenomena in nanofluid coolants with special focus on automotive applications. *Renewable and Sustainable Energy Reviews, 60*, 1615–1633.

26. N. A. C. Sidik, M. N. A. W. M. Yazid, & R. Mamat (2015). A review on the application of nanofluids in vehicle engine cooling system. *International Communications in Heat and Mass Transfer, 68*, 85–90.

27. B. M'hamed, N. A. C. Sidik, M. F. A. Akhbar, R. Mamat, & G. Najafi (2016). Experimental study on the thermal performance of MWCNT nano coolant in PeroduaKelisa 1000cc radiator system. *International Communications in Heat and Mass Transfer*, *76*, 156–161.
28. A. M. Hussein, R. A. Bakar, & K. Kadirgama (2014). Study of forced convection nanofluid heat transfer in the automotive cooling system. *Case Studies in Thermal Engineering*, *2*, 50–61.
29. S. U. S. Choi (1995). Enhancing thermal conductivity of fluids with nanoparticles ASME FED, vol. 231, p. 99–105.
30. Gujar, J., Patil, S., & Sonawane, S. (2024). Review on the Encapsulation, Microencapsulation, and Nano-Encapsulation: Synthesis and Applications in the Process Industry for Corrosion Inhibition. *Current Nanoscience*, *20*(3), 314–327.
31. Malika, M., & Sonawane, S. (2024). A Review on the Application of Nanofluids in Enhanced Oil Recovery. *Current Nanoscience*, *20*(3), 328–338.
32. Gujar, J. G., Patil, S. S., & Sonawane, S. S. (2023). A Review on Nanofluids: Synthesis, Stability, and Uses in the Manufacturing Industry. *Current Nanomaterials*, *8*(4), 303–318.
33. Malika, M., Pargaonkar, A., & Sonawane, S. S. (2023). Application of emulsion nanofluid membrane for the removal of methylene blue dye: stability study. *Chemical Papers*, 1–11.
34. S. U. S. Choi, & J. A. Eastman (1995). Enhancing thermal conductivity of fluids with nanoparticles. *American Society of Mechanical Engineering*, pp. 99–105.
35. L. Chen, W. Yu, & H. Xie (2012). Enhanced thermal conductivity of nanofluids containing Ag/MWNT composites. *Powder Technology*, *231*, 18–20.
36. M. Farbod, & A. Ahangarpour (2016). Improved thermal conductivity of Ag decorated carbon nanotubes water based nanofluids. *Physics Letters: Section A*, *380*, 4044–4048.
37. V. Kumar, & J. Sarkar (2018). Two-phase numerical simulation of hybrid nanofluid heat transfer in minichannel heat sink and experimental validation. *International Communications in Heat and Mass Transfer*, *91*, 239–247.
38. R. Ramachandran, K. Ganesan, & L. Asirvatham (2016). The role of hybrid nanofluids in improving the thermal characteristics of screen mesh cylindrical heat pipes. *Thermal Science*, *20*, 2027–2035.
39. N. A. C. Sidik, M. N. A. W. M. Yazid, & R. Mamat (2017). Recent advancement of nanofluids in engine cooling system. *Renewable and Sustainable Energy Reviews*, *75*, 137–144.
40. S. Senthilraja, K. Vijayakumar, & R. Gangadevi (2015). A comparative study on thermal conductivity of Al$_2$O$_3$/water, CuO/water and Al$_2$O$_3$– CuO/water nanofluids. *Digital Journal of Nanomaterial and Biostructures*, *10*, 1449–1458.
41. M. Gollin, & D. Bjork (1996). *Comparative performance of ethylene glycol/water and propylene glycol/water coolants in automobile radiators* (No. 960372). SAE Technical Paper.
42. D. G. Subhedar, B. M. Ramani, & A. Gupta (2018). Experimental investigation of heat transfer potential of Al$_2$O$_3$/water-mono ethylene glycol nanofluids as a car radiator coolant. *Case Studies in Thermal Engineering*, *11*, 26–34.
43. F. E. Lockwood, Z. G. Zhang, T. R. Forbus, S. U. Choi, Y. Yang, & E. A. Grulke (2005). *The current development of nanofluid research* (No. 2005-01-1929). SAE Technical Paper.
44. S. K. Saripella, W. Yu, J. L. Routbort, & D. M. France (2007). *Effects of nanofluid coolant in a class 8 truck engine* (No. 2007-01-2141). SAE Technical Paper.

45. Y. Lu, & S. Pilla (2017). Nanotechnology applications in future automobiles (2010-01-1149).

46. R. S. Vajjha, D. K. Das, & P. K. Namburu (2010). Numerical study of fluid dynamic and heat transfer performance of Al_2O_3 and CuO nanofluids in the flat tubes of a radiator. *International Journal of Heat and Fluid Flow*, *31*(4), 613–621.

47. K. Y. Leong, R. Saidur, S. N. Kazi, & A. H. Mamun (2010). Performance investigation of an automotive car radiator operated with nanofluid-based coolants (nanofluid as a coolant in a radiator). *Applied Thermal Engineering*, *30*(17–18), 2685–2692.

48. S. M. Peyghambarzadeh, S. H. Hashemabadi, M. S. Jamnani, & S. M. Hoseini (2011). Improving the cooling performance of automobile radiator with Al_2O_3/water nanofluid. *Applied Thermal Engineering*, *31*(10), 1833–1838.

49. A. M. Hussein, R. A. Bakar, K. Kadirgama, & K. V. Sharma (2014). Heat transfer enhancement using nanofluids in an automotive cooling system. *International Communications in Heat and Mass Transfer*, *53*, 195–202.

50. S. S. Chougule, & S. K. Sahu (2014). Comparative study of cooling performance of automobile radiator using Al_2O_3-water and carbon nanotube-water nanofluid. *Journal of Nanotechnology in Engineering and Medicine*, *5*(1).

51. S. M. Peyghambarzadeh, S. H. Hashemabadi, M. Naraki, & Y. Vermahmoudi (2013). Experimental study of overall heat transfer coefficient in the application of dilute nanofluids in the car radiator. *Applied Thermal Engineering*, *52*(1), 8–16.

52. S. Ramalingam, R. Dhairiyasamy, & M. Govindasamy (2020). Assessment of heat transfer characteristics and system physiognomies using hybrid nanofluids in an automotive radiator. *Chemical Engineering and Processing-Process Intensification*, *150*, 107886.

53. S. K. Soylu, İ. Atmaca, M. Asiltürk, & A. Doğan (2019). Improving heat transfer performance of an automobile radiator using Cu and Ag doped TiO_2 based nanofluids. *Applied Thermal Engineering*, *157*, 113743.

54. V. Kumar, & R. R. Sahoo (2021). Exergy and energy performance for wavy fin radiator with a new coolant of various shape nanoparticle-based hybrid nanofluids. *Journal of Thermal Analysis and Calorimetry*, *143*(6), 3911–3922.

55. R. R. Sahoo, & J. Sarkar (2017). Heat transfer performance characteristics of hybrid nanofluids as coolant in louvered fin automotive radiator. *Heat and Mass Transfer*, *53*(6), 1923–1931.

56. X. Li, H. Wang, & B. Luo (2021). The thermophysical properties and enhanced heat transfer performance of SiC-MWCNTs hybrid nanofluids for car radiator system. *Colloids and Surfaces A: Physicochemical and Engineering Aspects*, *612*, 125968.

57. F. Abbas, H. M. Ali, M. Shaban, M. M. Janjua, T. R. Shah, M. H. Doranehgard, … & F. Farukh (2021). Towards convective heat transfer optimization in aluminum tube automotive radiators: Potential assessment of novel Fe_2O_3-TiO_2/water hybrid nanofluid. *Journal of the Taiwan Institute of Chemical Engineers*, *124*, 424–436.

58. S-C. Tzeng, C-W. Lin, & K. D. Huang (2005). Heat transfer enhancement of nanofluids in rotary blade coupling of four-wheel-drive vehicles. *ActaMech*, *179*, 11–23.

59. K.-J. Zhang, D. Wang, F.-J. Hou, W.-H. Jiang, F-R. Wang, J. Li, G.-J. Liu, & W.-X. Zhang (2007). Characteristic and experiment study of HDD engine coolants. *NeiranjiGongcheng/Chinese Internal. Combust Engine Eng.*, *28*, 75–78.

60. S. K. Saripella, W. Yu, J. L. Routbort, D. M. France, & Rizwan-Uddin (2007). Effects of nanofluid coolant in a class 8 truck engine. *SAE Technical Papers*, art. No. 2007-01-2141.

61. M. Ali, A. M. El-Leathy, & Z. Al-Sofyany (2014). The effect of nanofluid concentration on the cooling system of vehicles radiator. *AdvMechEng.*, art. no. 962510.

62. M. Kole, & T. K. Dey (2010). Experimental investigation on the thermal conductivity

and viscosity of engine coolant based alumina nanofluids, In AIP conference proceedings, vol. 1249, pp. 120–124.

63. M. Kole, & T. K. Dey (2010). Thermal conductivity and viscosity of Al_2O_3 nanofluid based on car engine coolant. *Journal of Physics D: Applied Physics*, *43*, art. 315501.

64. M. M. Elias, I. M. Mahbubul, R. Saidur, M. R. Sohel, I. M. Shahrul, S. S. Khaleduzzaman, & S. Sadeghipour (2014). Experimental investigation on the thermo-physical properties of Al2O3 nanoparticles suspended in car radiator coolant. *International Communications in Heat and Mass Transfer*, *54*, 48–53.

65. Thakur, P. P., Sonawane, S. S., & Mohammed, H. A. (2023). Recent Trends in Applications of Nanofluids for Effective Utilization of Solar Energy. *Current Nanoscience*, *19*(2), 170–185.

66. Hakke, V. S., Gaikwad, R. W., Warade, A. R., Sonawane, S. H., Boczkaj, G., Sonawane, S. S., & Sapkal, V. S. (2023). Artificial neural network prophecy of ion exchange process for Cu (II) eradication from acid mine drainage. *International Journal of Environmental Science and Technology*, 1–12.

67. Malika, M., Jhadav, P. G., Parate, V. R., & Sonawane, S. S. (2023). Synthesis of magnetite nanoparticle from potato peel extract: its nanofluid applications and life cycle analysis. *Chemical Papers*, *77*(2), 1081–1094.

68. Choudhary, M., Singh, D., Jain, S. K., Sonawane, S. R. S., Singh, D., Devnani, G. L., & Srivastava, K. (2023). Kinetics modeling & comparative examine on thermal degradation of alkali treated Crotalaria juncea fiber using model fitting method. *Journal of the Indian Chemical Society*, 100918.

4 Applications of Nanofluids in the Boiling Processes

4.1 INTRODUCTION

In flow boiling heat transfer fluid is moved across the surface and heat transfer is driven by the flow. The flow boiling heat transfer concept is used by many industries for heat exchange systems. Boiling is a more efficient heat transfer phenomenon compared with others. Boiling is a phase-changing phenomenon from the liquid phase to the vapour phase. Boiling is the type of vaporization that occurs throughout the liquid. The boiling of a moving stream inside the channel can be referred to as flow boiling. Over the last few years, many investigators have attempted to understand the mechanism of flow boiling. Many investigators tried to understand and explain this phenomenon and they performed many experiments and simulations. Nanoparticles have the potential to enhance the thermal efficiency of basic fluid (HTC and CHF). Nanofluids can be prepared from nanometer size solid metals (like gold, silver, iron and copper), metals oxide (like CuO, Al_2O_3 and SiO_2) nonmetals and carbide to base fluid (like oil, ethylene glycol and water). When we add nanoparticles, it affects both surface characteristics and thermophysical properties.

Over the last few years, many researchers have been dealing with the flow boiling heat transfer which has grown at a very rapid rate because of its wide industrial and technological use. There are many research papers and articles available all over the world, which are related to this topic.

Flow boiling heat transfer is being used in many industrial applications such as boiler tubes, evaporators and cooling of rectors in a nuclear power plant. Nanofluids are new types of thermal fluids. It is a very new concept that came into existence in the 1990s. Flow boiling is a good method of transferring high heat fluxes. Flow boiling heat transfer can be limited by critical heat flux (CHF). By dispersing a nanometer sized solid particle (<100 nm) into base liquids such as water, engine oil and ethylene glycol nanofluids can be made. Nanoparticles enhance the thermal properties of the base fluids. Abdul Kaggwa [1] demonstrated that nanoparticles can't be well dispersed in their base fluids, surfactants are being used for proper dispersion of nanoparticles to ensure a high rate of heat transfer. This chapter discusses the recent studies focusing on the flow boiling heat transfer using nanofluids. Our main focus is on parameters like the heat transfer coefficient (HTC) and the critical heat flux (CHF) of nanofluids. Figure 4.1 represents the boiling mechanism of nanofluids for different heat flux values.

DOI: 10.1201/9781003404767-4

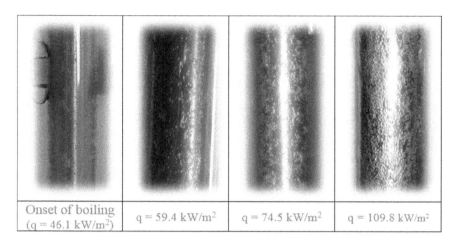

| Onset of boiling (q = 46.1 kW/m^2) | q = 59.4 kW/m^2 | q = 74.5 kW/m^2 | q = 109.8 kW/m^2 |

FIGURE 4.1   Different heat flux values for different kinds of nanofluids [2].

4.2   RECENT ADVANCES IN THE NANOFLUIDS APPLICATION IN THE BOILING PROCESS

K. Dolatiasl and Y. Bakshan [3] have done a simulation of subcooled flow boiling to predict Critical Heat Flux for water-Al$_2$O$_3$ nanofluids flows inside the tube. They got the result that with an increase in the temperature of fluid entering the tube, the Critical Heat Flux difference at different mass fluxes decreases. They used the Euler-Euler model to show the relationship of phases.

S.M. Peyghambarzadeh and A. Bayt [2] investigated HTC by using the reduced quadratic model in forced convection model in forced convection and subcooled flow boiling to CuO-water nanofluids and compared results (experimental and theoretical).

Dadhich and Prajapati [4] worked on (base fluid water, Al$_2$O$_3$, TiO$_2$ nanoparticle) nanofluids against DI water. They listed Critical Heat Flux for different heat flux, mass flux and concentrations. An artificial neural network (ANN) model was proposed to predict the Heat Transfer Coefficient for flow boiling of nanofluids. All values resulting from the ANN model were cross-checked with experimental values and found that values were very similar. Augusto and Ribatski [5] did an experimental investigation to evaluate CHF. The experiment was done for Al$_2$O$_3$, SiO$_2$ and base fluid DI-water. They tried to find how CHF and mass flux are related. Images of the CHF location and end of the tube are shown in Figure 4.2. Sarafraz and Khalil [6] performed an experimental study on the subcooled flow boiling heat transfer and pressure drop characteristics with carbon-terminal 66 nanofluids. Forced convective and nucleate boiling heat transfer regimes identified under flow boiling heat transfer.

E. Abedini and S. Niaz [7] showed that different parameters affect HTC and by considering all those parameters they presented four models to predict the critical heat flux of the subcooled nanofluids flow. The nanoparticles used are Al$_2$O$_3$, diamond, zinc oxide and graphene oxide. The model was suited with <21% MAE.

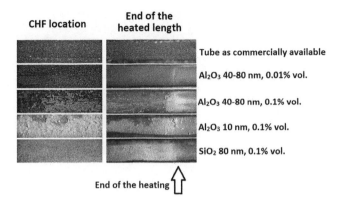

FIGURE 4.2 Images of the test section observed through the electron microscope [5].

Hamidi and Kianoush [8] came up with results that subcooled flow boiling condition lubrication force near the wall decreases the vapor volume fraction and liquid velocity, this helps to move bubbles from near the wall to the central regions. Bubble departure frequency, bubble departure diameter and nucleation site density (NSD) decrease in the base fluid after adding nanoparticles into it. They also proposed two correlations to calculate HTC and compared the results with experimental results and found MAE is <10%.

Mohammad and Mahdi [9] in their work used acetone-zinc bromide with graphene nanoparticles for experimental study. To reach different results computational fluid dynamics simulations were achieved. The experimental results show that by adding 4.5% (by volume) graphene thermal conductivity increased by 4.5% which helped to achieve a higher heat flow rate. Figure 4.3 represents the CFD simulation results obtained for the acetone vapors in the acetone/water nanofluids system.

4.3 EFFECT OF NANOFLUIDS ON HTC AND CHF

For the boiling heat transfer phenomenon, the upper limit of the heat transfer rate is called the CHF or CHF is the thermal limit of a phenomenon in which a phase change occurs during heating, which results in a reduction in the heat transfer rate. The thermal conductivity of nanofluids depends on the volume fraction of the nanoparticle, particle shape and size.

E. Abedini [10] reported that in subcooled flow boiling (Alumina water nanofluids), by decreasing in mass flow rate heat transfer rate can be increased or decreased; this depends on the effect of latent heat transport and forced convection. They used a two-phase mixture model. Majid and Hossein [11] proved the role of suspending nanoparticles in increasing density, velocity and temperature profile; by statistical data they showed that increasing the number of nanoparticles is not always economical for practical application.

Hashemi [12] showed that the HTC of MWCNTs-GA-water nanofluids increased with applied heat. The HTC of nanofluids was greater than that of pure water, even at

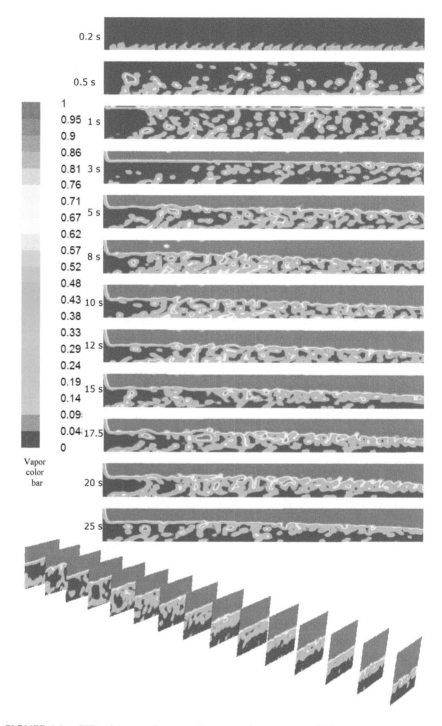

FIGURE 4.3   CFD of vapor of acetone in acetone/water system [9].

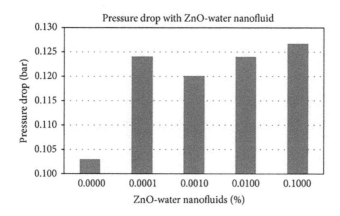

FIGURE 4.4 Variation of pressure drop for the nanoparticle concentration [13].

relatively low heat flux. The maximum HTC enhancement of the MWCNTs-GA-water nanofluids was about 4.3% at 0.005% concentration. The pressure drop of nanofluids was increasing with mass flux. Shankar [13] reported that by adding ZnO into water, HTC increased to 126% with a high-pressure drop and large increase in surface roughness. The reported results of pressure drop variation for the nanoparticle concentration are shown in Figure 4.4. The results reported in the literature are compared in Table 4.1.

4.4 PARAMETERS AFFECTING THE NANOFLUIDS PERFORMANCE

4.4.1 BUBBLE DYNAMICS

Soleimani and Sattari [26] showed that gas-phase bubble movements can cause disturbance in the thermal boundary layer near microchannel walls and will enhance heat transfer in comparison with single-phase flow. The local HTC increases after the starting of nucleate boiling. It is observed that the wall temperature distribution is directly affected by bubble movement and nucleation. These numeric results are shown in Figure 4.5.

Tao and Zhao [27] proved that with the increase in bubble distance velocity of bubble fusion decreases which leads to a decrease in heat transfer. Hamidi and Kianoush [8] came up with the result that subcooled flow boiling condition lubrication force decreases the liquid velocity and vapor volume fraction values near the wall. Bubble departure frequency, bubble departure diameter and nucleation site density mainly depend on Reynolds number. Bubble departure diameter and nucleation site density start decreasing in the base fluid after adding nanoparticles into it. They also proposed two correlations to calculate HTC and compared the results with experimental results and found that MEA is <10%.

Salari [19] showed that roughness of surface, fouling resistance and static contact angle are very important parameters for enhancing HTC. With a lowering contact angle, n. of nucleation sites decreases which results in weaker bubble transport and lower HTC.

TABLE 4.1

Literature Review of Nanofluids Application in Boiling Process

References	Nanofluids Type/Concentration	Heat/Mass Flux/Pressure	HTC Results	CHF Results
Sajjad [14]	Fe_3O_4-water	77,000 W/m² 270 kg/m². s	increased	–
A. Zangeneh [2]	CuO-water	–	increased	–
Hamidi and Kianoush [8]	Al_2O_3 and TiO_2 in R-113/0.5%, 1% and 3% (by volume)	–	increased	–
Kuanghan Deng [15]	SiC, Graphite in water/0.1% (by volume)	413-965 kg/m². s /48–289 KW/m²/ 0.21–0.89MPa	increased	increased
Ahn & Lee [16]	water in a micro-structured Zirlo tube	300 to 1500 kg/m². S at atmospheric pressure	–	increased
Shankar [13]	ZnO-water/0.0001 to 0.1% (by volume)	400 kg/m². s /0-400 KW/m²/1–2.5 bar	increased	–
E. Abedin [10]	Al_2O_3-water	–	increased	–
Augusto [17]	Al_2O_3/SiO_2-water/ 0.001, 0.01 and 0.1% (by volume)	200 to 600 kg/m². s /350 KW/m²	increased	increased
Lee & Park [18]	SiC/Al_2O_3-water/ 0.01% (by volume)	low flow and low-pressure condition	–	increased
Salari [19]	alumina-water/1% (by volume)	600 kg/m² s /50 KW/m² /at normal pressure	increased	–
Lee and Kam [20]	magnetite-water	high exit quantity and LFD condition	–	negligible
Mukherjee [21]	Al_2O_3-water/0.01, 0.05, 0.1, 0.5, 1% (by volume)	25 °C–65 °C	increased	–
Sarafraz [22]	CuO in water/ethylene glycol/0.5, 1, 1.5% (by volume)	353–1,059 kg/m². s /174 KW/m²/353 and 363 K	increased	–
you and Kim [23]	Al_2O_3DI water	540 to 670 KW/m²/60 °C	–	increased
Zhang [24]	Graphene oxide (GO)-water/0 to 0.05% (by weight)	–	increased	–
Sung and Chain [25]	graphene-Cu	1–20 W/m² 240–480Kg/m². s	increased	–

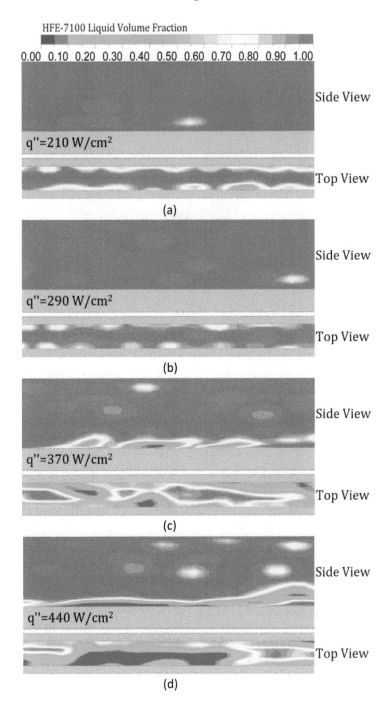

FIGURE 4.5 Top view and side view of various flux with a mass flow rate of 5 g/s for a) 210 W/cm², b) 290 W/cm², c) 370 W/cm², and d) 440 W/cm² [26].

(a)   (b)

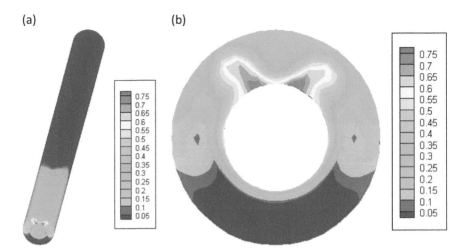

FIGURE 4.6 (a) Vapor volume fraction along the cross-section area of geometry (b) Vapor volume fraction along the vertical axis of geometry [28].

4.4.2 FLOW DIRECTION

Masoud and Hamid [28] studied the flow of boiling HTC for nanofluids (Al_2O_3/TiO_2-water) in different directions of flow. Due to different flow directions, buoyancy force came into the picture which affects vapor volume fraction resulting final effect on HTC. These numeric results are shown in Figure 4.6.

Mukherjee and Mishra [21] experimented with alumina water nanofluids in the horizontal tube and found a 94% enhancement in overall thermal performance compared to pure water. Also compared results for the horizontal and vertical tubes.

4.4.3 EFFECT OF FOULING ON HEAT TRANSFER

Sarfaraz and Hormoz [22] did an experimental investigation on heat transfer conditions and parameters and found that heat and mass flux are directly proportional to flow boiling HTC and fouling resistance. When the concentration of nanoparticle rises, flow boiling HTC decreases, while fouling resistance rises. The rate of fouling and thickness is very dependent on time.

Heat transfer during phase change comprises different sub-phenomena like bubble departure, bubble dynamics, nucleation site density, waiting time of bubble growth, detachment frequency of bubbles, Evaporation, condensation, etc. The presence of nanoparticles as the solid changes phase in the boiling process raises this complication due to the interaction between the phases. To tackle this complexity many new models and correlations were proposed in recent years. The effect of variation of heat flux on the bubble formation around the heater can be observed in Figure 4.7.

Ahn and Kim [29] examined a saturated water alumina nanofluids pool boiling in 10 mm diameter at atmospheric pressure conditions near CHF. By comparing the

HF= 50W/m²   HF=73kW/m²   HF=92kW/m²   HF=120kW/m²   HF=130kW/m²

FIGURE 4.7 Effect of the value of heat flux on bubble formation [22].

value for water-Al_2O_3 nanofluids results, a measurable increase in wall temperature can be seen after attaining critical heat flux. At the time of nanofluids boiling, a bend in the boiling curve was noticed at a high heat flux.

4.5 FLOW BOILING HEAT TRANSFER CORRELATION

E. Abedini and S. Niaz [7] developed a few co-relations for the flow boiling. Mohammad and Walker [30] performed computational fluid dynamics simulations to study the effect of different nanoparticle concentrations, boiler temperature and fluid velocity on boiling and phase change behavior. Four phases of vapour acetone, liquid acetone, zinc bromide solution and solid nanoparticles cloud were treated here. ZnO nanoparticles are considered as a cloud-like phase in the mixture. Figure 4.8 represents the distribution of all phases in the boiling process. Sarfraz and Kamal [31] also developed various co-relations for understanding the mechanism of the flow boiling process.

Ziba and Mehrzad [32] modelled the nanofluids boiling as a three-phase model. Three phases, vapour, water and nanoparticles were taken into consideration to model the nanofluids flow boiling. With the help of the Eulerian 2-fluid model, the subcooled flow boiling of water was simulated. The results from the experiment and three-phase model are agreeable. The results of the simulation show:

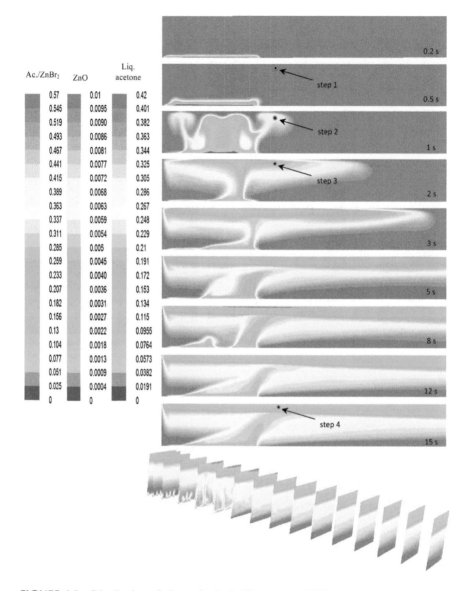

FIGURE 4.8 Distribution of phases in the boiling process [30].

1. The inverse relation is between vapor void fraction and inlet velocity. The conventional heat transfer can be enhanced by enhancing the fluid velocity.
2. Proportional relation between vapor volume fraction and wall heat flux, vapor volume fraction rises with wall heat flux.

Various co-relations reported in the literature are shown in Table 4.2.

TABLE 4.2
Flow Boiling Heat Transfer Correlation

References	Correlation For	Mass and Heat Flux	Working Pressure/Temperature	Nanoparticle Used	Mean Absolute Error	Model
E. ABedini and S. Niaz [7]	CHF	$100 < G$ [kg/m².s] < 2500	100, 400, 800 KPa	Alumina, Diamond, ZnO, Graphene-Oxide	9%	Binary model
Dadhich [4]	HTC	143.2 kg/m²– 1014 kg/m².s	—	Al_2O_3 & TiO_2	±10.5% and ±10%	ANN Model
E. Abedini [3]	CHF	—	—	alumina	<10%	Euler Model
S.M. Peyghambarzadeh and A. Bayt [2]	HTC	—	—	CuO-water	—	reduced quadratic model
Hamidi and Kianoush [8]	HTC		330, 333, 335 K	Al_2O_3 and TiO_2/R-113	<10%	Binary model
Mohammad and Walker [30]	CHF	—		zinc oxide with acetone	<6%	—
Sarfraz [22]	HTC	353–1059 kg/m² s/ 174 kW/m²	353 and 363 K	CuO in water/ethylene glycol	<11%	Chen correlation
Wang and Wu [33]	CHF	98.9–348.4 kg/m² s	13.5–35.9 C/ 0.40–0.89 MPa	Al_2O_3, AlN	—	correlation
Ayyaz Ketan [23]	CHF	24.1–806.5 Kg/m². s	—	silica water	12.1%	correlation

4.6 CONCLUSIONS

This chapter reviews the latest studies related to flow boiling by using nanofluids. The first part talked about the topic; the second part is related to what has been happening in the last few years related to the topic. The third part discussed how different parameters affect flow boiling using nanofluids. Over the past few years development related to the flow boiling using nanofluids has been unique. CHF, subcooled flow boiling and HTC are the topics related to flow boiling using nanofluids on which recent studies are mainly based. Only a few correlations were available regarding bubble dynamics. A more systematic experimental study can be conducted to learn about more bubble parameters by using nanofluids.

REFERENCES

1. A. Kaggwaa, J. K. Carson, M. Atkins, & M. Walmsley (2019). The effect of surfactants on viscosity and stability of activated carbon, alumina and copper oxide nanofluids. www.materialstoday.com/proceedings
2. A. Zangeneh, S. Peyghambarzadeh, A. Bayat, & A. Vatani (2020). Application of general multilevel factorial design approach in forced convection and subcooled flow boiling heat transfer to CuO/water nanofluids. 10.1016/j.molliq.2020.113502
3. K. Dolatias, Y. Bakhshan, E. Abedini1, & S. Niazi (2019). Numerical investigation of critical heat flux in subcooled flow boiling of nanofluids. 10.1007/s10973-019-08616-
4. M. Dadhich, O. S. Prajapati, & N. Rohatgi (2019). Flow boiling heat transfer analysis of Al2O3 and TiO2 nanofluids in horizontal tube using artificial neural network (ANN). 10.1007/s10973-019-08674-y
5. F. J. do Nascimento, T. A. Moreira, & G. Ribatski (2019). Flow boiling critical heat flux of DI-water and nanofluids inside smooth and nano porous round microchannels. www.elsevier.com/locate/ijhmt
6. M. Sarafraz, & A. T. K. Abad (2019). Statistical and experimental investigation on flow boiling heat transfer to carbon nanotube-therminol nanofluids. www.elsevier.com/locate/physa
7. K. DolatiAsl, Y. Bakhshan, E. Abedini, & S. Niazi (April 2019). Correlations for estimating critical heat flux (CHF) of nanofluid flow boiling. www.elsevier.com/locate/ijhmt
8. E. Abedini, A. M. Karachi, R. H. Jahromi, & K. DolatiAsl (2020). Numerical investigation of flow boiling of refrigerant-based nanofluids and proposing correlations for heat transfer. DOI: 10.1177/0954408920929771
9. H. I. Mohammed, D. Giddingsb, G. S. Walker, P. Talebizadehsardaric, & J. M. Mahdie (2020). Thermal behaviour of the flow boiling of a complex nanofluid in a rectangular channel: An experimental and numerical study. www.elsevier.com/locate/ichmt
10. E. Abedini, A. Behzadmehr, S. Sarvari, & S. Mansouri (2012). Numerical investigation of subcooled flow boiling of a nanofluid. www.elsevier.com/locate/ijts
11. M. Zarringhalam, H. Ahmadi-Danesh-Ashtiani, D. Toghraie, & R. Fazaeli (2019). The effects of suspending Copper nanoparticles into Argon base nanofluid inside a microchannel under boiling flow condition by using of molecular dynamic simulation. 10.1016/j.molliq.2019.111474
12. M. Hashemi, & S. H. Noie (2017). Study of flow boiling heat transfer characteristics of critical heat flux using carbon nanotubes and water nanofluid. DOI 10.1007/s10973-017-6661-1

13. O. S. Prajapati, & N. Rohatgi (2014). Flow boiling heat transfer enhancement by using ZnO-water nanofluids. 10.1155/2014/890316
14. Y. Wang, K. Deng, J. Wu, G. Su, & S. Qiu (2018). The characteristics and correlation of nanofluid flow boiling critical heat flux. 10.1016/j.ijheatmasstransfer.2018.01.118
15. Y. Wanga, K. Dengc, J. M. Wu, G. Su, & S. Qiuc (2020). A mechanism of heat transfer enhancement or deterioration of nanofluid flow boiling heat transfer. www.elsevier.com/locate/hmt
16. H. S. Ahn, S. H. Kang, C. Lee, J. Kim, & M. H. Kim (2012). The effect of liquid spreading due to micro-structures of flow boiling critical heat flux. www.elsevier.com/locate/ijmulflow
17. T. A. Moreira, F. J. do Nascimento, & G. Ribatski (2017). An investigation of the effect of nanoparticle composition and dimension on the heat transfer coefficient during flow boiling of aqueous nanofluids in small diameter channels (1.1 mm). www.elsevier.com/locate/etfs
18. S. W. Lee, S. D. Park, S. Kang, S. M. Kim, H. Seo, & D. W. Lee (2013). Critical heat flux enhancement in flow boiling of Al 2O 3 and SiC nanofluids under low pressure and low flow conditions. https://www.researchgate.net/publication/264078567
19. E. Salari, M. Peyghambarzadeh, M. M. Sarafraz, & F. Hormozi (2016). Boiling heat transfer of alumina nano-fluids: Role of nanoparticle deposition on the boiling heat transfer Coefficient. 10.3311/PPch.9324
20. T. Lee, D. H. Kam, J. H. Lee, & Y. H. Jeong (2014). Effects of two-phase flow conditions on flow boiling CHF enhancement of magnetite-water nanofluids. www.elsevier.com/locate/ijhmt
21. S. Mukherjee, S. Jana, P. C. Mishra, P. Chaudhuri, & S. Chakrabarty (2020). Experimental investigation on thermo-physical properties and subcooled flow boiling performance of Al2O3/water nanofluids in a horizontal tube. http://www.elsevier.com/locate/ijts
22. M. Sarfraz, & F. Hormoz (2014). Convective boiling and particulate fouling of stabilized CuO-ethylene glycol nanofluids inside the annular heat exchanger. 10.1016/j.icheatmasstransfer.2014.02.019
23. A. Siddique, K. Sakalkale, S. K. Saha, & A. Agrawal (2020). Investigation of flow distribution and effect on critical heat flux in multiple parellal microchannel flow boiling. 10.1007/s00231-020-02972-0
24. C. Zhang, L. Zhang, H. Xu, D. Wang, & B. Ye (2017). Investigation of flow boiling performance and the resulting surface deposition of graphene oxide nanofluid in microchannels. 10.1016/j.expthermflusci.2017.03.028
25. H. I. Mohammed, D. Giddings, G. S. Walker, P. Talebizadehsardari, & J. M. Mahdi (2020). Thermal behaviour of the flow boiling of a complex nanofluid in a rectangular channel: An experimental and numerical study. 10.1016/j.icheatmasstransfer.2020.104773
26. A. Soleimani, A. M. Sattari, & P. Hanafizadeh (2020). Thermal analysis of a microchannel heat sink cooled by two-phase flow boiling of Al2O3 HFE-7100 nanofluid. 10.1016/j.tsep.2020.100693
27. S. Yao, T. Huang, K. Zhao, J. Zeng, & S. Wang (2019). Simulation of flow boiling of nanofluid in tube based on lattice Boltzmann model. 10.2298/TSCI160817006Y
28. M. Afrand, E. Abedini, & H. Teimouri (2010). Experimental investigation and simulation of flow boiling of nanofluids in different flow directions. www.elsevier.com/locate/physe
29. H. S. Ahn, & M. H. Kim (2013). The boiling phenomenon of alumina nanofluid near critical heat flux. www.elsevier.com/locate/ijhmt
30. H. I. Mohammed, D. Giddings, & G. S. Walker (2018). CFD simulation of a concentrated salt nanofluid flow boiling in a rectangular tube. www.elsevier.com/locate/ijhmt

31. M. M. Sarfraz, F. Hormoz, & M. Kamalgharibi (2014). Sedimentation and convective boiling heat transfer of CuO-water/ethylene glycol nanofluids. DOI 10.1007/s00231-014-1336-y

32. Z. Valizadeh, & M. Shams (2015). Numerical investigation of water-based nanofluid subcooled flow boiling by three-phase Euler–Euler, Euler–Lagrange approach. DOI 10.1007/s00231-015-1675-3

33. Y. Wang, K. Deng, J. Wu, G. Su, & S. Qiu (2018). The characteristics and correlation of nanofluid flow boiling critical heat flux. 10.1016/j.ijheatmasstransfer.2018.01.118

5 Applications of Nanofluids for the CO_2 Absorption Process

5.1 INTRODUCTION

With the increase in industrialization, researchers are also working on novel methods of carbon capture and regenerating the captured carbon. The use of membranes, adsorption of carbon dioxide on suitable solid beds or absorption by various solvents are a few strategies that are conventionally used for this purpose. Various strategies are being developed to improve the performance of these methods. Absorption of carbon dioxide in liquid solvents is the most widely used method among all these methods. It is comparatively easier to regenerate the medium used in the absorption for carbon capture than other methods. Absorption of carbon dioxide can be physical absorption or chemical absorption. Physical absorption does not involve any chemicals, thus, this method is comparatively less efficient but, can be more easily maintained. Whereas using chemical absorption for carbon capture has some limitations. With the change in the operating parameter, the risk of side-reaction is always there. Corrosion and degradation of equipment may occur due to this chemical reactivity. Thus, for better safety, physical absorption is preferred in many cases for CO_2 absorption.

Nanofluids are generally used for heat transfer enhancement, but studies showed that nanofluids can increase the mass transfer rate. The use of nanofluids for the absorption of CO_2 and other gases has increased in recent years. Thus, it is important to understand the mechanism of mass transfer enhancement by nanofluids and develop the most suitable nanofluids for mass transfer operations. Jung et al. [1] developed a methanol based nanofluids system with nearly 8% CO_2 absorption enhancement. Ghosh et al. [2] have developed a nanofluid with the graphitic carbon nitride nano-sheets. Using these nanoparticles the 19.78 mmol/g CO_2 is absorbed in the base fluid. But amine-based nanofluids have shown better results than other nanofluids Arshadi et al. [3] used DEA as base fluid and Fe_3O_4 as a nanoparticle. This system reported a CO_2 enhancement of 70%. Taheri et al. [4] used Al_2O_3 instead of Fe_3O_4 and the same DEA as base fluid, the CO_2 absorption is 33% was observed.

Sumin et al. [5] performed the experiments for the CO_2 absorption by nanofluids of alumina and carbon nanotubes. After comparing the results of these nanofluids with each other, the carbon nanotube-based nanofluids reported better CO_2 absorption capacity. This may be due to the morphology of carbon nanotubes, the result of these experimentations is shown in following Figure 5.1. Here stirring speed and solid concentration are maintained constant.

DOI: 10.1201/9781003404767-5

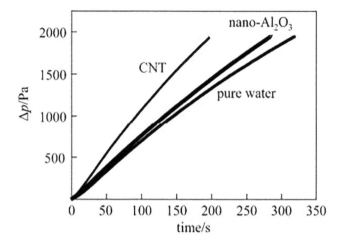

FIGURE 5.1 CO_2 absorption capacity for water and CNT and alumina-based nanofluids [5].

Lee et al. [6] compared the two metal oxide-based nanofluids i.e., alumina based nanofluids and silica-based nanofluids. At 0.01% nanoparticle concentration, the alumina-based nanofluids absorbed 4.5% CO_2, while silica-based nanofluids absorbed 5.6% CO_2 at 20 °C. Researchers have attempted to develop the mechanism of enhancement and general rules for the nanofluids absorption capacity. Zhang et al. [7] reported the advances in the nanofluids application for the CO_2 absorption. The role of stability of nanofluids is highlighted by the author. Liu et al. [8] used 1-dimethylamino-2-propanol as a working fluid for the CO_2 absorption. The absorption is mainly a chemical absorption. This is dependent on the concentration of 1DMA2P. Gibbs-Helmholtz equation is used to calculate heat of CO_2 absorption i.e., heat of reaction. This value is −31.67 kJ/mol. Rehman et al. [9] used a hollow membrane collector for CO_2 absorption operation. The CO_2 absorption can be effectively increased by using the MWCNT-based membranes for this purpose. Other strategies like the use of hydrophilic nanomaterials are also used by researchers. Li et al. [10] synthesized a nanofluids system with the hydrophilic oleate-modified Fe_3O_4 nanoparticles. These nanofluids have recorded an increase in the CO_2 absorption. Compared to other amines, methyl di-ethyl amine (MDEA) has better CO_2 absorption capacity. Irani et al. [11] added graphene oxide as a nanoparticle in MDEA, absorption of CO_2 is 9.3%. Haghtalab et al. [12] compared the performance of silica-based nanofluids and zinc oxide-based nanofluids and reported that the zinc oxide-based nanofluids have better results than the alumina-based nanofluids. Figure 5.2 represents the performance of ZnO nanoparticles at 5 °C.

Fly ash based nanofluids are also important for the sustainable development of the system. Fly ash is made up of silica and alumina, as we have already seen, silica and alumina are good choices for mass transfer enhancement. Nabipour et al. [13] synthesized the fly ash based nanofluids with sulfinyl-M as a base fluid. These nanofluids increased the efficiency of mass transfer by 7%. Lu et al. [14] synthesized the nanofluids of four different nanoparticles. These nanoparticles are ferrous oxide, carbon nanotubes, silica oxide and aluminum oxide. The author

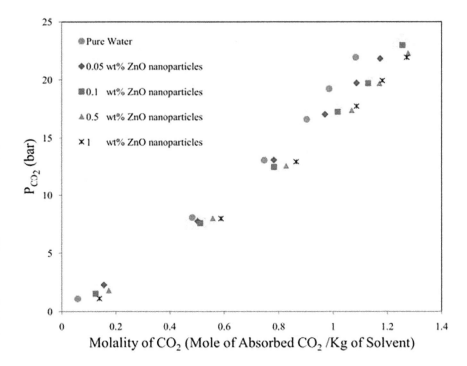

FIGURE 5.2   Effect of ZnO nanoparticle concentration on CO_2 absorption [12].

reported that, for distilled water as a base fluid, ferrous oxide has comparatively better results than other nanoparticles. Ferrous oxide has reported the enhancement of 53% CO_2 absorption at 0.15 wt%. Similarly, various other researchers have developed various nanofluids based on the amine as the base fluid and various nanoparticles [15–18]. It is important to develop a suitable design and engineer the nanofluids for that particular application. The longer stability of nanofluids and easy regeneration are challenges for the efficient nanofluids based absorption system. [19].

5.2 PRINCIPLE

The increase in the absorption due to nanofluids is explained by three different mechanisms. Grazing/Shuttle effect, breaking of boundary layer due to hydrodynamics of the solid-liquid, and bubble breaking. These mechanisms are schematically represented in Figure 5.3 [20].

5.2.1 GRAZING (SHUTTLE EFFECT)

Nanoparticles present in the nanofluids move in Brownian motion. These nanoparticles are always in random motion. These nanoparticles adhere to the bubble surface and increase the bubble velocity within the column. Because of this random movement bubbles move in bulk of liquid without any external force. This movement gives the gas bubble more contact area with the bulk of the liquid. Thus,

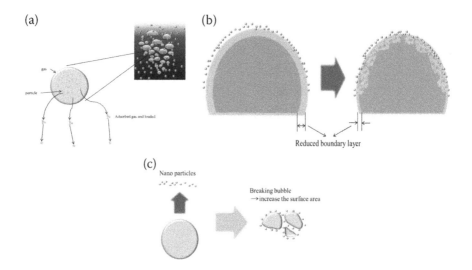

FIGURE 5.3 Different mechanisms of CO_2 absorption (a) grazing effect, (b) hydro-dynamic effect, and (c) weakening of bubble [23].

the mass transfer increases. The presence of nanoparticles also contributes to the localized turbulence in the system. This turbulence also contributes to the mass transfer enchantment [21,22]. Sometimes the bunch of nanoparticles may carry a single bubble in the bulk of the liquid. This bubble sometimes bursts into smaller-sized bubbles. This again increases the surface area for mass transfer. This effect is called the shuttle effect because the nanoparticles work as a shuttle and carry the gas bubble from one place to another. This phenomenon is also called as grazing effect. The presence of nanoparticles on the bubble surface weakens the bubble and the bubble gets divided into various smaller-sized bubbles. This phenomenon is represented in Figure 5.3(a).

5.2.2 WEAKENING OF THE LIQUID-GAS BOUNDARY RESISTANCE

Another very important reason behind the increase in mass transfer is a reduction in the boundary layer between gas and liquid. This effect is schematically represented in Figure 5.3(b). Nanoparticles adhere to the bubble surface. This weakens the layer on the bubble surface. Thus, the bubble breaks down into several smaller bubbles. This smaller bubble offers more surface area for mass transfer. These results are reported by various researchers, Kluytmanset et al. [24] reported that the diffusion layer was weakened due to the presence of nanoparticles on the bubble surface. [25].

5.2.3 BREAKING OF THE BUBBLES

An increase in the mass transfer due to nanofluids depends not only on the grazing effect or adhesion of nanoparticles but on both mechanisms. Both mechanisms contribute to the increase in mass transfer. Apart from this, various other design changes can be made to increase the mass transfer rate [26–28]. Nanofluid systems are hindered by external forces to improve the turbulence in the system [29]. This phenomenon is represented in Figure 5.3c.

5.3 AMINE-BASED NANOFLUIDS SYSTEM

Amines are preferred as a solvent for the CO_2 absorption operation. In Figure 5.4 the chemical structure of various amines is given. Di-ethanol amine and methyl di-ethanol amines are the most widely used solvents in the industry. The reaction mechanism of amines and carbon dioxide is studied by Caplo [30] and Danckwerts [31]. These reaction schemes are shown in Eq. 5.1 to Eq. 5.7. These schemes contain the reaction between water and CO_2 also. But, absorption is mainly due to the amines. Thus, the reaction between water and CO_2 can be neglected.

$$H_2O \leftrightarrow H^+ + OH^- \; H^+ + OH^- \tag{5.1}$$

$$CO_2 + H_2O \leftrightarrow H^+ + HCO_3^- \tag{5.2}$$

$$CO_2 + OH^- \leftrightarrow HCO_3^- \tag{5.3}$$

$$HCO_3^- \leftrightarrow CO_3^{2-} \tag{5.4}$$

$$CO_2 + R_1NH_2 \leftrightarrow R_1NH_2^+COO^- \tag{5.5}$$

$$R_1NH_2^+COO^- + R_1NH_2 \leftrightarrow R_1NH_3^+ + R_1NCOO^- \tag{5.6}$$

$$R_1NCOO^- + H_2O \leftrightarrow R_1NH_2 + HCO_3^- \tag{5.7}$$

Industrial growth and development in the various industries led to more carbon dioxide emissions. Industries are hesitant to implement the methods required to reduce CO_2 emissions due to the high cost of such methods. Out of the total carbon dioxide emissions; the power generation process contributes most than any other

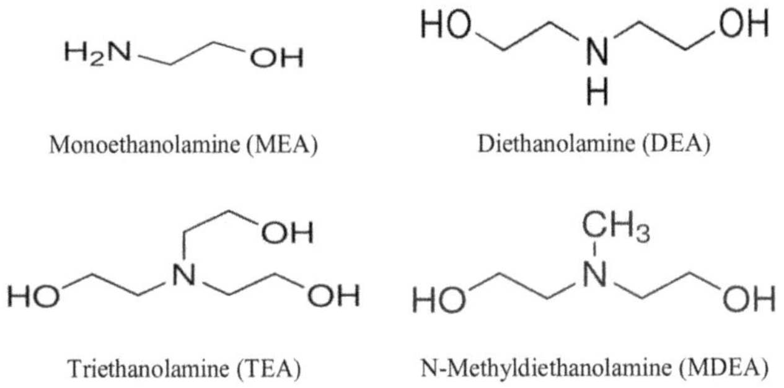

Monoethanolamine (MEA)   Diethanolamine (DEA)

Triethanolamine (TEA)   N-Methyldiethanolamine (MDEA)

FIGURE 5.4 Structures of different amines [29].

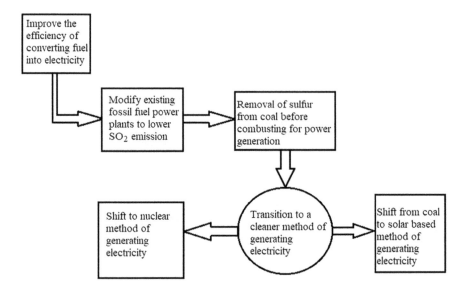

FIGURE 5.5 Strategy for the reduction in CO_2 [32].

process. The power generation process contributes more than 38% to global CO_2 emissions. Then vehicles and industry contribute second highest than other industrial processes. The transportation industry contributes more than 26% to the global CO_2 emission. The rest of the industrial processes contribute more than 24% of the world's CO_2 emission. These all industrial processes are collectively responsible for 90% of the CO_2 emission in the atmosphere. This emission of CO_2 into the atmosphere has a very bad impact on the environment and human health. Thus, to reduce the CO_2 emission various techniques are being explored by various researchers. These techniques are represented in Figure 5.5.

As we have discussed, absorption of carbon dioxide is as in reaction scheme 5.1 to reaction scheme 5.7. These reaction schemes can be written down as overall reaction form as given in equations 5.8 and 5.9.

$$2RNH_2 + CO_2 \leftrightarrow RNH_3^+ + RNHCOO^- \tag{5.8}$$

$$RNH_2 + H_2O + CO_2 \leftrightarrow RNH_3^+ + HCO_3^- \tag{5.9}$$

As we have seen in Table 5.1, fossil fuel combustion is one of the major reasons behind CO_2 emission. Generally, the mixture of water and amines is used in industries for carbon capture and storage. The general process flow diagram of the carbon dioxide capture in the industries is shown in Figure 5.6.

Amine-based nanofluids are preferred, because of the following reasons:

- Compared to other solvents, amines can absorb very little concentration of CO_2.
- The cost of amines is also very cheap (approximately 5 USD per litre).

TABLE 5.1

Nanoparticles and Setups Used for the Carbon Dioxide Absorption Study [34]

Adsorbent	Enhancement Factor	Testing Setup	The Concentration of CO_2 (g CO_2/g)	References
CNT	1.45	Hollow fiber	0.026	35
Fe_2O_4	1.3	membrane	0.023	
SiO_2	1.04	(HFMC)	0.021	
Al_2O_3	1.1		0.018	
ZnO	1.48	Hollow fiber	0.007	36
TiO_2	1.26	membrane	0.006	
MWCNT	1.18	(HFMC)	0.005	
Fe_3O_4	1.23	Bubble column	–	3
$Fe_3O_4@SiO_2$	1.1		–	
$Fe_3O_4@ SiO_2\text{-}NH_2$	1.45		–	
$Fe_3O_4@ SiO_2\text{-}SNH_2$	1.52		–	

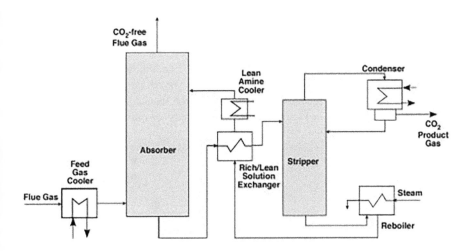

FIGURE 5.6 Schematic representation of amine-based carbon dioxide absorption [33].

- If the contact time is given enough, the amines can absorb 98% of CO_2.
- It is easy to operate with amine-based nanofluids.

But, the following are a few drawbacks of the amine-based nanofluids system:

- Amines-based nanofluids require a high amount of energy.
- Amines generally have a low boiling point. The thermophoresis effect can lead to a decrease in the absorption capacity of nanofluids.

- Amines can lead to the corrosion of material of construction and side reactions of amine can result in the formation of salts and other unwanted components.
- The solvent can be degraded due to a low boiling point if the hot flue gases are used for the carbon dioxide absorption.
- Generally, gases coming out of industry contain fly ash, which may result in foaming.

Various studies have shown the absorption of carbon dioxide with various nanoparticles in the nanofluids. These studies are tabulated in Table 5.1.

5.4 PARAMETERS AFFECTING THE MASS TRANSFER

5.4.1 EFFECT OF NANOFLUID TYPE

As we have seen in Chapter 1, the viscosity of the nanofluids increases with the increase in the nanoparticle concentration. The same results are reported in the carbon dioxide absorption studies. Shah et al. [37] performed experimentations with the polydimethylsiloxane (PDMS) based nanofluids. In this study, CuO nanoparticles were used. This CuO/PDMS nanofluids system had a viscosity of 2.28 cp. Change in the absorption efficiency is also reported for different nanofluid systems. Kim et al. [38] performed experiments for the K_2CO_3/piperazine solutions with different concentrations of silica as a nanoparticle. For small concentrations of silica i.e., 0.021 wt%, nearly 12% absorption of carbon dioxide is achieved. Absorption achieved by silica is higher than alumina-based nanofluids. Pineda et al. [39], conducted an experimental study with alumina based nanofluids. The results obtained from alumina based nanofluids are better than silica-based nanofluids. These results are shown in Figure 5.7. A similar study was also conducted by other

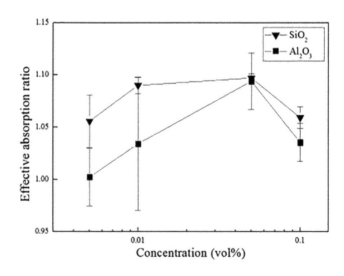

FIGURE 5.7 Comparative performance of silica and alumina nanofluids [39].

researchers like Taheri et al. [40]. Jiang et al. [26] performed various experiments with different nanoparticles like Al_2O_3, MgO, SiO_2, and TiO_2. From the various nanofluids, it is evident that TiO_2 gives better results than other nanoparticles. Thus, the author recommended the nanofluids system of TiO_2/MDEA. This system has performed better compared to other methods.

5.4.2 EFFECT OF NANOPARTICLE SIZE

Kim et al. [38] performed experimental studies with different nanoparticle sizes and observed that there is not much difference in the CO_2 absorption capacity of nanofluids. This is because the bubble size doesn't get affected by the nanoparticle size. The size of the nanoparticle has affected the heat transfer rate but, no enhancement is reported for the mass transfer operation [41]. Nagy et al. [42] studied the mass transfer rate of carbon dioxide absorption with various nanoparticles with mathematical models and proposed that, at smaller nanoparticle sizes, more mass transfer rate is observed. Thus, the more research and detailed analysis of this area is needed. The results reported in this study are shown in Figure 5.8.

5.4.3 EFFECT OF NANOPARTICLE LOADING

Lee et al. [43] performed various experiments with the various nanofluids with different nanoparticle concentrations. The carbon dioxide capture with the various concentrations of alumina and silica nanoparticles is represented in Figure 5.9. As we can see, the trend of results for both nanoparticles is similar. The absorption efficiency of the nanofluids increases with the increase in the nanoparticle concentration. After the 0.01% by volume nanoparticle concentration, the absorption efficiency of nanofluids decreases. This is because overcrowding of nanoparticles at

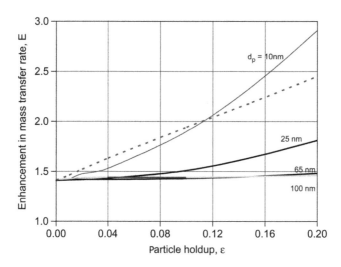

FIGURE 5.8 Effect of nanoparticle size on the mass transfer rate [42].

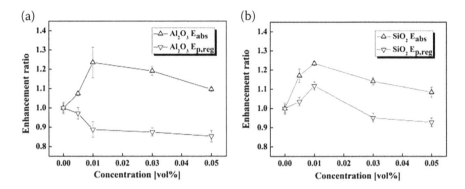

FIGURE 5.9 Effect of nanoparticle concentration enhancement ratio: (a) alumina-based nanofluids and (b) silica-based nanofluids [43].

gas-liquid interphase results in a barrier for mass transfer. This decreases the nanofluid's performance. Similar results were reported by other researchers. Taheri et al. [40] used alumina and silica to study the effect of nanoparticle concentration on the CO_2 absorption efficiency. The absorption efficiency increased up to 0.05 wt% of nanoparticle concentration. Regeneration of nanoparticles also decreases with the addition of nanoparticles for alumina-based nanofluids and increases for silica-based nanofluids. However, after a certain point, the regeneration efficiency of nanofluids does not increase. Darabi et al. [44] also reported the same results with nanofluids of carbon nanotubes and silica. The absorption efficiency increased initially, up to the nanoparticle concentration of 0.2 wt% and then reduced after this concentration. This trend can be observed in all the studies reported.

5.5   CASE STUDY OF FLY ASH BASED NANOFLUIDS

The fly ash used for the nanofluids synthesis is collected from the Koradi power plant, Nagpur, (MS), India. The surface area of fly ash is between 550 and 880 m^2/g. The quantitative analysis of the fly ash component is done by X-ray fluorescence (Make: Malvern Panalyticals). The composition of fly ash on a mass basis was SiO_2 (54.72%), Al_2O_3 (34.69%), Fe_2O_3 (2.34.36%), TiO_2 (1.03%), MgO (2.12%), CaO (1.6%), etc. The size reduction of fly ash is done using a Spex-8000 ball mill. Fly ash size is reduced to 80 nm. Figure 5.10 represents the dynamic light scattering equipment results for the size of the nanoparticles. Distilled water, 0.1 N NaOH, and 0.1 M di-ethanolamine are used as base fluids and 0.25% of nanoparticle concentration. Sodium dodecyl sulfate (SDS) dispersant is added. Synthesis of nanofluid is done either by a single step or a two-step process, the latter being more effective and applicable from a practical point of view. In the two-step technique, the nanoparticles are dispersed in the base solution, the particles start to agglomerate and settle down reducing the efficiency of the nanofluids. To prevent the agglomeration of the particles various physical and chemical methods are employed, ultra-sonication being the most effective and widely used method for

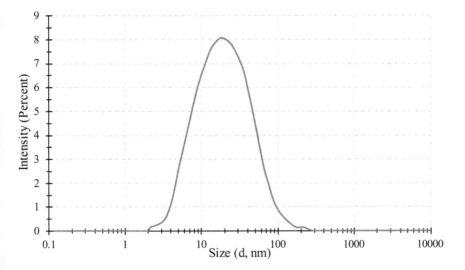

FIGURE 5.10　DLS result of fly ash nanoparticles.

stabilizing the nanofluids. The number of fly ash nanoparticles that need to be added to the respective base fluid is calculated using equation 5.10.

$$\Phi \times 100 = \frac{W_p/\rho_p}{W_p/\rho_p + W_{bf}/\rho_{bf}} \times 100 \tag{5.10}$$

Ultrasonication is used to improve the dispersion of nanofluids. Probe-type sonication equipment (E-chromTech, Taiwan, power rating: 800 W, frequency: 20 kHz) is used for 20–30 minutes. The zeta potential of the prepared nanofluids is measured using the dynamic light scattering equipment (Malvern ZETASIZER NANO ZS). Figure 5.10 shows the zeta potential values of the various combinations of nanofluids.

Most of the researchers have explored the experimental approach for the carbon dioxide absorption study using nanofluids. In the introduction of this chapter, we have seen that various studies are being explored by the experimental methods. Thus, in this section with the help of fly ash based nanofluids, we have discussed the experimental parameters and their effect on the absorption of carbon dioxide. Generally, aqueous forms of various chemicals like NaOH, and Di-ethyl amines are used as base fluid. To understand the experimental approach for the nanofluids based systems, we used fly ash nanoparticles. The two-step approach is used for the fly ash synthesis. Here 20 micron sized fly ash is reduced by 60 hours of ball milling. These nanoparticles are then added to the base fluid and a stable suspension of nanofluids is prepared by the suitable ultra-sonication of suitable sonication time or by using surfactant. The stability of nanofluids is evaluated by the dynamic light scattering equipment by measuring the zeta potential of nanofluids. Generally, a glass column is used for the experimentation. Here, we also have a bubble column

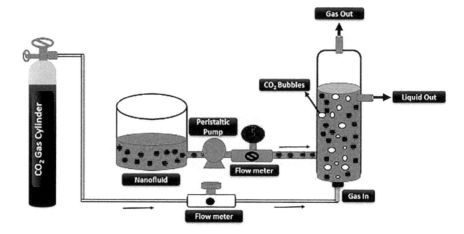

FIGURE 5.11   Schematic representation of CO_2 absorption setup.

of height 45 cm for our study. The schematic representation of this experimentation is given in figure 5.11.

This column is initially filled with the working fluid. Carbon dioxide gas is passed from the bottom of the column. The even distribution of gas is ensured by the sparger. This even distribution of sparger is important for the better efficiency of the column. The design changes can be made in the column. Various researchers have used different packing, and various flow patterns are also used for study. The pictures of the column are taken with a high-speed camera. To measure the approximate average diameter of the bubble. The concentration of carbon dioxide can be measured by a line carbon dioxide analyzer or the titration method can be used. The exiting CO_2 concentration is then measured with the help of a titration method.

The sample of the bulk of the liquid is collected after some time, once equilibrium is achieved. The following formulas are used for the calculations.

The volume of CO_2 absorbed per unit of time is given by equation 5.11

$$\Delta V = Q_{in} - Q_{out} (C_{in} - C_{out})/(1 - C_{out}) \tag{5.11}$$

Where

Q_{in} = inlet gas volume flow rate
Q_{out} = outlet gas volume flow rate
C_{in} = initial CO_2 volume fraction
C_{out}= residual CO_2 volume fraction.

The solvents used for the absorption of CO_2 is di-ethanolamine, water and NaOH. Reactions involved in the chemical absorption of CO_2 in di-ethanolamine are represented by equation 5.12 and the chemical absorption of CO_2 in water is represented by equation 5.13.

$$DEAH^+ + CO_2 \leftrightarrow DEACOO^- + H^+ \tag{5.12}$$

$$CO_2 + H_2O \leftrightarrow HCO_3^- + H^+ \tag{5.13}$$

To carry out the titration, NaOH is added to the samples collected from the amine-based nanofluids and water-based nanofluids. No need for NaOH addition in NaOH-based nanofluids. Reactions involved in CO_2 analysis by titration are as follows:

$$HCO_3^- + NaOH \rightarrow Na_2CO_3 + H_2O \tag{5.14}$$

The titration process is carried out with the help of HCl. This process is represented by equation 5.15 and equation 5.16

$$Na_2CO_3 + HCl \rightarrow NaHCO_3 + H_2O \tag{5.15}$$

$$NaHCO_3 + HCl \rightarrow NaCl + H_2O \tag{5.16}$$

The first reaction marks the first endpoint of the titration when carried out with phenolphthalein indicator while the second reaction marks the second endpoint when titrated using methyl orange as the indicator. These points can be seen with a digital pH meter.

The difference between the HCl used in the second endpoint and the first endpoint is the total amount of HCl used in neutralizing the absorbed CO_2.

The concentration of the consumed CO_2 is determined by equation 5.17:

$$M_1 V_1 = M_2 V_2 \tag{5.17}$$

M_1 = Molarity of HCl, M_2 = Molarity of CO_2,
V_1 = Volume of HCl used up, V_2 = Volume of nanofluid containing CO_2

5.5.1 ENHANCEMENT OF CO₂ CAPTURE USING NANOFLUIDS

The improved absorption of CO_2 using nanofluids for the base fluid is represented by an enhancement factor.

Percentage enhancement in CO_2 absorbed is determined from equation 5.18

$$E_s\,(\%) = 100 \times (CO_{2\ absorbed\ by\ nanofluid} - CO_{2\ absorbed\ by\ base\ fluid})/$$
$$CO_{2\ absorbed\ by\ base\ fluid} \tag{5.18}$$

5.6 RESULT AND DISCUSSION OF RESULTS OBTAINED IN THE CASE STUDY

5.6.1 STABILITY STUDY

Zeta potential is an important measure of the stability of nanofluids. The stability of nanofluids is represented in Figure 5.12. Amines-based nanofluids have more zeta potential values than NaOH. This better stability contributes to the enhancement of

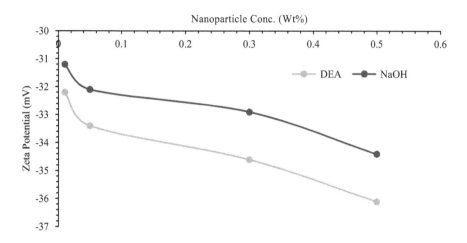

FIGURE 5.12   Comparison of zeta potential values for NaOH- and DEA-based nanofluids.

the CO_2 absorption study. Thus, less agglomeration of nanoparticles can be seen in the amine-based nanofluids. Less agglomeration of nanoparticles leads to the smaller size of gas bubbles in the bulk of the liquid. An increase in the concentration of amine or NaOH leads to an increase in the zeta potential of nanofluids.

From Figure 5.12, it is evident that di-ethanol amine-based nanofluid is more stable. DEA-based nanofluid has a stability of −36 mV at a nanoparticle concentration of 0.5 wt%. Nanoparticle concentration above 0.5 wt% value has no significant impact on the zeta potential of nanofluids. Maximum value achieved by the NaOH-based nanofluid is −34 mV at 0.5 wt%. A similar trend can be observed for the nanoparticles concentration below 5 wt%.

5.6.2   EFFECT OF THE FLY ASH CONCENTRATION ON CO_2 ABSORPTION

The experimental results of different fly ash nanoparticle concentrations in nanofluids on CO_2 absorption are shown in Figure 5.13. The result shows that the amount of CO_2 increases with the increase in the nanoparticle concentration. The grazing effect of nanoparticles can be seen as more dominant in the higher nanoparticle concentration. The stability of nanofluids after 0.5 wt% nanoparticle concentration remains the same. The increase in the nanoparticle concentration has no significant impact on the CO_2 absorption after 5 wt% of nanoparticle concentration. The chances of agglomeration of nanoparticles are also high at higher nanoparticle concentrations. An increase in the nanoparticles decreases the hydrodynamic boundary between the gas bubble and the liquid bulk. Thus, an increase in mass transfer is observed in nanofluids. For the water-based nanofluids at a 40 lpm flow rate, the observed nanofluids absorption is 35% and it is 55% higher than the absorption observed in the water alone as a solvent for the same flow rate. A similar trend is observed for the di-ethanol amine-based nanofluids and NaOH-based nanofluids. The absorption efficiency of the fly ash/ NaOH nanofluid at a 40 lpm flow rate is 58% higher than 0.1 M NaOH as a solvent. Similarly, the absorption efficiency of fly ash/amine-based nanofluids is 65% more

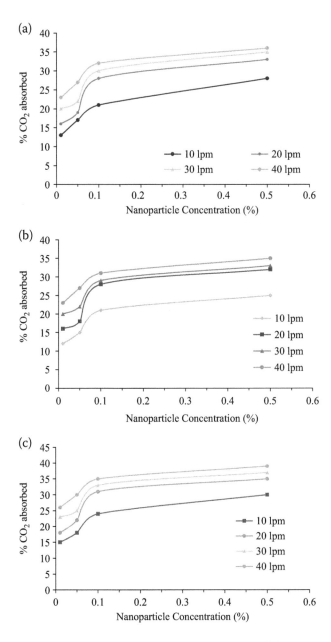

FIGURE 5.13 Effect of the nanofluids flow rate of various nanofluids on CO₂ absorption: (a) water-based nanofluids, (b) NaOH-based nanofluids, (c) DEA-based nanofluids.

than the Di-ethanol amine alone as a solvent. Even at the concentration of 0.1 wt% of fly ash nanoparticles, the absorption efficiency is more than 40% in all three cases. Thus, a further increase in the nanoparticle concentration has a less significant impact on CO₂ absorption.

5.6.3 Flow Rate Study

Figure 5.13 represents the comparative analysis of the effect flow rates of various nanofluids on CO_2 absorbed. The performance of water-based nanofluid, NaOH-based nanofluid and DEA-based nanofluid is studied for flow rates of 10 lpm, 20 lpm, 30 lpm and 40 lpm. Approximately 46% more CO_2 gas is absorbed in water-based nanofluids with 0.5 wt% concentration at 40 lpm than the 10 lpm flow rate. A higher flow rate offers more turbulence in the system and thus, a higher mass transfer can be seen. At low flow rates, the residence time is longer, but it is observed that the turbulent interaction of the bubble and liquid bulk is more dominant in the CO_2 absorption process than the residence time of the bubble. The increase in CO_2 absorption is similar at low nanoparticle concentrations. The increase in the CO_2 absorption at 0.1 wt% of nanoparticle concentration is 44%. Similarly, the amount of CO_2 absorption is 45% to 50% for fly ash/NaOH-based nanofluids and fly ash/DEA nanofluids for all four flow rates. This shows that the enhancement of CO_2 absorption is similar for all fly ash nanoparticle concentrations. A flow rate below 10 lpm was not possible due to the back pressure of the liquid column. It was not possible with the available experimental setup. Similarly, data for a flow rate above 40 lpm was not possible due to operational difficulties due to the high flow rate.

5.6.4 Desorption Study

Solar thermal energy is used to separate the absorbed CO_2 from the fly ash/nanofluids system. The average solar irradiation in Nagpur, Maharashtra state, India is 1,266.52 W/sq.m. The incident radiation is allowed to pass through nanofluids in a microchannel. Microchannel offers a better heat transfer area. The size of the micro-channel is 800 microns. The suitable residence time can offer better heat transfer and thus, ultimately the better desorption of CO_2 from the nanofluids. From Figure 5.14 it is evident that the desorption of nanofluids is dependent on the flow rate of nanofluids. The higher the flow rate, the less amount of CO_2 will be desorbed. In the case of the fly ash/water system, the amount of CO_2 desorbed is maximum at a 1 lpm flow rate and decreases with an increase in the flow rate for all three nanofluids systems. Nearly, 70% of CO_2 gets desorbed at high nanoparticle concentrations. This is because, at high temperatures, the specific heat of nanofluids decreases, and CO_2 trapped in nanofluids gets released. Nanofluids with low nanoparticle concentrations have comparatively high specific heat. Thus, CO_2 desorption is difficult in the low concentration of fly ash nanoparticles.

5.6.5 Bubble Dynamics

The behaviour of gas bubbles in an aqueous medium is an important parameter to study the mass transfer and heat transfer of a gas-liquid system. Thus, in a fly ash based nanofluids system, the bubble dynamics are dependent on the temperature of nanofluids, the flow rate of solvent, the flow rate of the gas phase, and the geometry

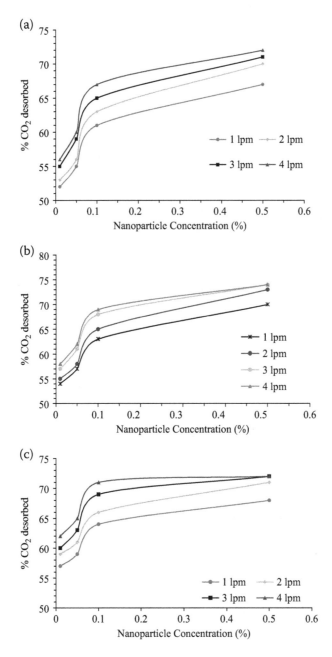

FIGURE 5.14 Desorption study for various nanofluids: (a) water-based nanofluids, (b) NaOH-based nanofluids, (c) DEA-based nanofluids.

of the column. During the experimentation, various images of the column are captured. These captured images are useful to determine the bubble dynamics and mechanism of CO_2 absorption in gas bulk. These images are also important to develop mathematical models and numeric studies. These images are used to calculate the mean bubble diameter and cycle time using the ImageJ software. These results are also useful to develop the mathematical correlation between other flow patterns and column geometry and dimensions. Figure 5.15 represents the real-time images of bubbles in the liquid bulk of fly ash/water nanofluids and fly ash/DEA nanofluids.

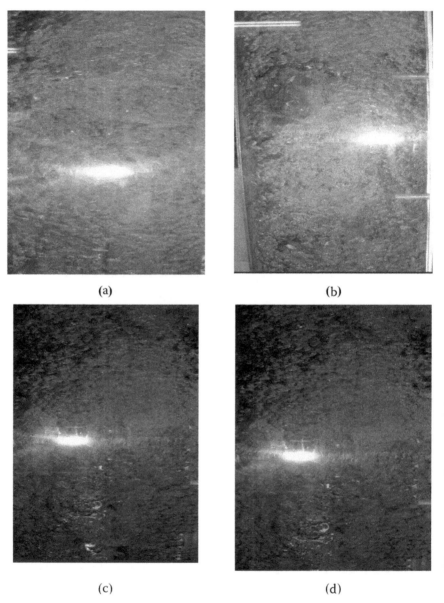

(a)   (b)

(c)   (d)

FIGURE 5.15 Real-time images of the absorption column. (a) 0.1 wt% fly ash/water nanofluids; (b) 0.5 wt% fly ash/water nanofluids; (c) 0.1 wt% fly ash/DEA nanofluids; (d) 0.5 wt% fly ash/DEA nanofluids.

5.6.5.1   Average Bubble Diameter

The average bubble diameter for various concentrations of fly ash/water nanofluids and fly ash/DEA nanofluids at various concentrations is represented in Figure 5.16. An increase in the nanoparticle concentration decreases the average diameter of gas bubbles in the liquid bulk. The amount of nanoparticles increases the grazing effect. Nanoparticles are always in random motion. Because of this random motion, the nano/micro-sized gas bubbles also travel with the nanoparticle movement in the liquid bulk and increase the CO_2 absorption. The increased nanoparticles at the gas surface decrease the hydrodynamic boundary layer of gas and liquid. This increases the mass transfer of CO_2 gas in the nanofluids. From Figure 5.16, it is clear that the increased fly ash nanoparticles in the fly ash/water nanofluids and fly ash/DEA nanofluids

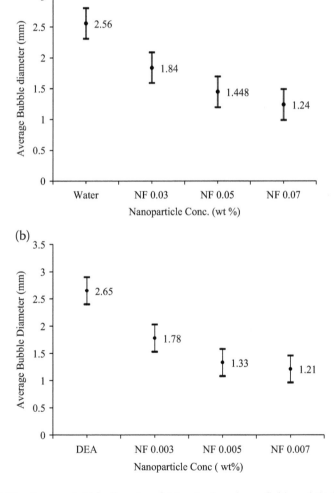

FIGURE 5.16   Average bubble diameter of (a) water-based nanofluids and (b) DEA-based nanofluids.

decrease the gas bubble size and this is a significant impact on the mass transfer rate of nanofluids. The average size of the bubble in water as a solvent is 2.56 mm while, the average size of the bubble in fly ash/water nanofluids of 0.07 wt% nanoparticles is 1.24 mm and this is a 51% reduction in the size of the bubble. Similarly, a 54.33% reduction in bubble size is observed in the nanofluids system.

5.6.5.2 Cycle Times

The cycle time is the total time spent by a bubble in the column. The gas flow is kept at 1×10^{-5} m^3/sec. The videos are captured by the slow motion mode of the camera. The comparative values of fly ash/water and fly ash/DEA nanofluids are shown in Figure 5.17. The cycle time of fly ash/water nanofluids is minimum at

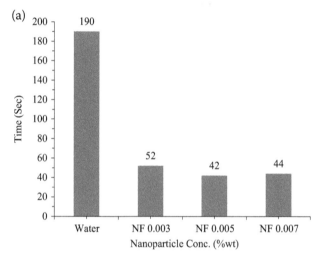

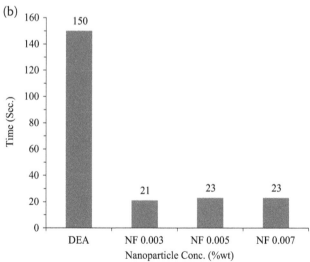

FIGURE 5.17 Total time of bubble cycle for (a) water-based nanofluids and (b) DEA-based nanofluids.

0.5 wt% nanoparticle concentration. For 0.7 wt% nanoparticle concentration, the cycle time again increases. This is due to the increase in nanoparticles leading to agglomeration of nanoparticles on the bubbles and this agglomeration often generates the cluster of bubbles in the liquid bulk. Thus, the higher nanoparticle concentrations should be avoided for better absorption efficiency. Similarly, fly ash/DEA nanofluids at 0.5 wt% and 0.7 wt% of fly ash nanoparticles have the same cycle time and this cycle time is more than the 0.3 wt% fly ash/DEA nanofluids. These results are shown in Figure 5.17.

5.6.5.3 Bubble Departure Frequency (f)

The time required to grow and dissipate the gas bubble from the sparger is also an important parameter of study. Equation 5.19 is useful to calculate the bubble departure frequency. This parameter calculates the bubble dissipation rate from the sparger.

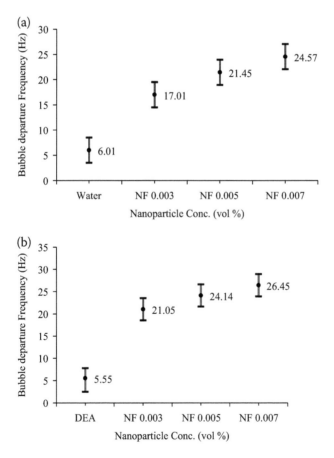

FIGURE 5.18 Bubble departure frequency for (a) fly ash/water-based nanofluids and (b) fly ash/DEA-based nanofluids.

$$f = \frac{1}{t_w + t_g} \qquad (5.19)$$

In this equation, t_w is the time required to wait for the next bubble to start to generate. t_g is the time required to grow the bubble and dissipate in the bulk of the liquid. Figure 5.18 represents the bubble departure frequency for the fly ash/water nanofluids system and fly ash/DEA system. From Figure 5.18, it is evident that the frequency of bubble dissipation is 4 times in the water-based nanofluids and 4.75 times in DEA-based nanofluids.

5.7 CHALLENGES AND FUTURE PERSPECTIVES

The need for efficient carbon capture techniques is increasing due to increasing demand from industries around the world. In section 5.1, various reasons for the increasing carbon dioxide in the atmosphere are discussed. The role of nanofluids in these novel methods is very crucial.

- However, the mechanism and stability for a longer period is still unclear. However, researchers are developing hybrid nanofluids for more carbon dioxide. It is important to develop mathematical relations for the carbon dioxide absorption and operational parameters.
- The stability of nanofluids is a very important issue of the nanofluids application that needs to be addressed.
- Thermophoresis effect prevents the use of nanofluids for the hot flue gas. The durability of the nanofluids is also a challenge to develop an efficient nanofluids system.
- The increase in the viscosity is also a matter of concern for various researchers, as it increases the operational cost of the process.
- The cost of nanoparticles and nanofluids synthesis also needs to be considered before choosing the solvent for carbon dioxide absorption.
- Higher concentrations of nanoparticles in the nanofluids can become the reason for the wear and tear of mechanical parts.
- The desorption of carbon dioxide is not discussed in detail.

Thus, it is important to address the above issues to develop more efficient nanofluids based carbon dioxide absorption and desorption systems.

5.8 CONCLUSION

Nanofluids have proved themselves as a good candidate for the novel solvent for efficient mass transfer. Various research studies are reported in this chapter to understand the effect of operational parameters on carbon dioxide absorption. Various attempts are being made to improve the stability of nanofluids. The modification of nanoparticles to stabilize the nanoparticle in the working fluid is one such approach. The challenges and future perspectives of this study are reported in the chapter. In this chapter, the role of nanofluids in the CO_2 absorption study is discussed in detail. The industrial solvents and recent advances in the application of nanofluids in carbon dioxide absorption are also discussed in detail. The numerical approaches for the modeling and simulation are briefly described. The effect of various parameters on the absorption efficiency is given in detail. The experimental approach for the carbon dioxide study is reported. The effect of various operational parameters and the need for a stable nanofluid system is discussed in detail. The bubble dynamics and role of the grazing effect and hydrodynamic effect of nanoparticles on the bubble are discussed in detail.

A case study of CO_2 absorption is also discussed in this chapter in detail. In this study, fly ash/water nanofluids, fly ash/NaOH and fly ash/DEA nanofluids systems are studied. The zeta potential analysis, flow rate study, and effect of nanoparticle concentration on CO_2 absorption are studied in detail. The flow rate of 40 lpm had the same enhancement in CO_2 absorption for all three nanofluid systems. An enhancement of 45% is recorded for the fly ash/water nanofluids system and a 55% enhancement in the CO_2 absorption in the fly ash/amine nanofluids system. The 0.5 wt% nanoparticle concentration has shown the best results for all the nanofluid systems. The desorption of nanofluids is done by solar radiation and a micro-channel-based system. The detailed analysis of nanofluids is done by the bubble dynamics of the nanofluids system.

REFERENCES

1. J. Y. Jung, J. W. Lee, & Y. T. Kang (2012). CO_2 absorption characteristics of nanoparticle suspensions in methanol. *Journal of mechanical science and technology*, *26*(8), 2285–2290.
2. S. Ghosh, & S. Ramaprabhu (2017). High-pressure investigation of ionic functionalized graphitic carbon nitride nanostructures for CO_2 capture. *Journal of CO_2 Utilization, 21*, 89–99.
3. M. Arshadi, H. Taghvaei, M. K. Abdolmaleki, M. Lee, H. Eskandarloo, & A. Abbaspourrad (2019). Carbon dioxide absorption in water/nanofluid by a symmetric amine-based nanodendritic adsorbent. *Applied Energy, 242*, 1562–1572.
4. M. Taheri, A. Mohebbi, H. Hashemipour, & A. M. Rashidi (2016). Simultaneous absorption of carbon dioxide (CO_2) and hydrogen sulfide (H2S) from CO_2–H_2S–CH_4 gas mixture using amine-based nanofluids in a wetted wall column. *Journal of Natural Gas Science and Engineering, 28*, 410–417.
5. L. U. Sumin, X. I. N. G. Min, S. U. N. Yan, & D. O. N. G. Xiangjun (2013). Experimental and theoretical studies of CO_2 absorption enhancement by nano-Al_2O_3 and carbon nanotube particles. *Chinese Journal of Chemical Engineering, 21*(9), 983–990.
6. J. W. Lee, I. T. Pineda, J. H. Lee, & Y. T. Kang (2016). Combined CO_2 absorption/regeneration performance enhancement by using nanoabsorbents. *Applied Energy, 178*, 164–176.
7. Z. Zhang, J. Cai, F. Chen, H. Li, W. Zhang, & W. Qi (2018). Progress in enhancement of CO_2 absorption by nanofluids: A mini review of mechanisms and current status. *Renewable Energy, 118*, 527–535.
8. H. Liu, H. Gao, R. Idem, P. Tontiwachwuthikul, & Z. Liang (2017). Analysis of CO_2 solubility and absorption heat into 1-dimethylamino-2-propanol solution. *Chemical Engineering Science, 170*, 3–15.

9. Z. U. Rehman, N. Ghasem, M. Al-Marzouqi, & N. Abdullatif (2019). Enhancement of carbon dioxide absorption using nanofluids in hollow fiber membrane contactor. *Chinese Journal of Chemical Engineering*.

10. Q. Li, R. Zhang, D. Wu, Y. Huang, L. Zhao, D. Wang, ... & G. Ma (2016). Cell-nanoparticle assembly fabricated for CO_2 capture and in situ carbon conversion. *Journal of CO2 Utilization*, *13*, 17–23

11. V. Irani, A. Maleki, & A. Tavasoli (2019). CO_2 absorption enhancement in graphene-oxide/MDEA nanofluid. *Journal of Environmental Chemical Engineering*, 7(1), 102782.

12. A. Haghtalab, M. Mohammadi, & Z. Fakhroueian (2015). Absorption and solubility measurement of CO_2 in water-based ZnO and SiO_2 nanofluids. *Fluid Phase Equilibria*, *392*, 33–42.

13. M. Nabipour, P. Keshavarz, & S. Raeissi (2017). Experimental investigation on CO_2 absorption in Sulfinol-M based Fe_3O_4 and MWCNT nanofluids. *International Journal of Refrigeration*, *73*, 1–10.

14. Y. Lu, J. Yan, & E. Dahlquist (2008). Experimental investigation on CO_2 absorption using absorbent in hollow fiber membrane contactor.

15. F. Cao, H. Gao, H. Ling, Y. Huang, & Z. Liang (2020). Theoretical modeling of the mass transfer performance of CO_2 absorption into DEAB solution in hollow fiber membrane contactor. *J. Memb. Sci.*, *593*, 117439.

16. A. Marjani, A. TaghvaieNakhjiri, A. Adimi, H. FathinejadJirandehi, & S. Shirazian (2020). Modification of polyethersulfone membrane using MWCNT-NH_2 nanoparticles and its application in the separation of azeotropic solutions by means of pervaporation. *PLoS One*, *15*, e0236529.

17. H. Mohammadnejad, S. Liao, B. A. Marion, K. D. Pennell, & L. M. Abriola (2020). Development and validation of a two-stage kinetic sorption model for polymer and surfactant transport in porous media. *Environ. Sci. Technol.*, *54*, 4912–4921. Mohammadnejad, H., Marion, B., Kmetz, A., Johnston, K.P., Pennell, K., Abriola, L.J.E.S.N., 2021. Development and Experimental Evaluation of a Mathematical Model to Predict Polymer-Enhanced Nanoparticle Mobility in Heterogeneous Formations.

18. M. Pishnamazi, A. T. Nakhjiri, A. S. Taleghani, A. Marjani, A. Heydarinasab, & S. Shi-razian (2020). Computational investigation on the effect of [Bmim][BF4] ionic liquid addition to MEA alkanol amine absorbent for enhancing CO_2 mass transfer inside membranes. *J. Mol. Liq.*, *314*, 113635

19. A. Rosli, N. F. Shoparwe, A. L. Ahmad, S. C. Low, & J. K. Lim (2019). Dynamic modelling and experimental validation of CO_2 removal using hydrophobic membrane contactor with different types of absorbent. *Sep. Purif. Technol.*, *219*, 230–240.

20. Q. Sohaib, A. Muhammad, M. Younas, & M. Rezakazemi (2020). Modeling pre-combustion CO_2 capture with tubular membrane contactor using ionic liquids at elevated temperatures. *Sep. Purif. Technol.*, *241*, 116677.

21. R. L. Kars, R. J. Best, & A. A. H. Drinkenburg (1979). The sorption of propane in slurries of active carbon in water. *Chem. Eng. J.*, *17*, 201-210.

22. D. Brilman, & G. Versteeg (1998). A one-dimensional stationary heterogeneous mass transfer model for gas absorption in multiphase systems. *Chem. Eng. Process*, *37*, 471–488.

23. J. H. Kim, C. W. Jung, & Y. T. Kang (2014). Mass transfer enhancement during CO_2 absorption process in methanol/Al_2O_3 nanofluids. *Int. J. Heat. Mass Tran*, *76*, 484–491.

24. J. H. J. Kluytmans, B. G. M. van Wachem, B. F. M. Kuster, & J. C. Schouten (2003). Mass transfer in sparged and stirred reactors: influence of carbon particles and electrolyte. *Chem. Eng. Sci.*, *58*, 4719–4728.

25. V. S. J. Craig (2004). Bubble coalescence and specific-ion effects, Curr. Opin. *Colloid & Interface Sci.*, *9*, 178–184.

26. J. Jiang, B. Zhao, Y. Zhuo, & S. Wang (2014). Experimental study of CO$_2$ absorption in aqueous MEA and MDEA solutions enhanced by nanoparticles. *Int. J. Greene. Gas. Con*, *29*, 135–141.

27. J.-Y. Jung, J. W. Lee, & Y. T. Kang (2012). CO$_2$ absorption characteristics of nanoparticle suspensions in methanol. *J. Mech. Sci. Technol.*, *26*, 2285–2290.

28. M. Jeong, J. W. Lee, S. J. Lee, & Y. T. Kang (2017). Mass transfer performance enhancement by nanoemulsion absorbents during CO$_2$ absorption process. *Int. J. Heat. Mass Tran*, *108*, 680–690.

29. A. T. Nakhjiri, A. Heydarinasab, O. Bakhtiari, & T. Mohammadi (2018). Modeling and simulation of CO$_2$ separation from CO$_2$/CH$_4$ gaseous mixture using potassium glycinate, potassium argininate and sodium hydroxide liquid absorbents in the hollow fiber membrane contactor. *J. Environ. Chem. Eng.*, *6*, 1500–1511.

30. M. Caplow (1968). Kinetics of carbamate formation and breakdown. *Journal of the American Chemical Society*, *90*(24), 6795–6803.

31. P. V. Danckwerts (1979). The reaction of CO$_2$ with ethanolamines. *Chemical Engineering Science*, *34*(4), 443–446

32. K. O. Yoro, & M. O. Daramola (2020). CO$_2$ emission sources, greenhouse gases, and the global warming effect. In *Advances in carbon capture* (pp. 3–28). Woodhead Publishing.

33. B. Miller (2015). 8-Greenhouse gas-carbon dioxide emissions reduction technologies. *Fossil fuel emissions control technologies*, 367–438.

34. H. A. Salih, J. Pokhrel, D. Reinalda, I. AlNashf, M. Khaleel, L. F. Vega, … & M. A. Zahra (2021). Hybrid–Slurry/Nanofluid systems as alternative to conventional chemical absorption for carbon dioxide capture: A review. *International Journal of Greenhouse Gas Control*, *110*, 103415.

35. B. Rahmatmand, P. Keshavarz, & S. Ayatollahi (2016). Study of absorption enhancement of CO$_2$ by S$_i$O$_2$, Al$_2$O$_3$, CNT, and Fe$_3$O$_4$ nanoparticles in water and amine solutions. *Journal of Chemical & Engineering Data*, *61* (4), 1378–1387.

36. P. Zare, P. Keshavarz, & D. Mowla (2019). Membrane absorption coupling process for CO$_2$ capture: Application of water-based ZnO, T$_i$O$_2$, and multi-walled carbon nanotube nanofluids. *Energy Fuels*, *33*(2), 1392–1403

37. R. D. Shah (2009). Application of nanoparticle saturated injectant gases for EOR of heavy oils.

38. W.-G. Kim, H. U. Kang, K.-M. Jung, & S. H. Kim (2008). Synthesis of silica nanofluid and application to CO$_2$ absorption. *Sep. Sci. Technol.*, *43*, 3036e3055.

39. I. Torres Pineda, J. W. Lee, I. Jung, & Y. T. Kang (2012). CO$_2$ absorption enhancement by methanol-based Al$_2$O$_3$ and S$_i$O$_2$ nanofluids in a tray column absorber. *Int. J. Refrig.*, *35*, 1402e1409.

40. M. Taheri, A. Mohebbi, H. Hashemipour, & A. M. Rashidi (2016). Simultaneous absorption of carbon dioxide (CO$_2$) and hydrogen sulfide (H2S) from CO 2eH 2 SeCH 4 gas mixture using amine-based nanofluids in a wetted wall column. *J. Nat. Gas Sci. Eng.*, *28*, 410e417.

41. F. M. Ali, W. M. M. Yunus, & Z. A. Talib (2013). Study of the effect of particles size and volume fraction concentration on the thermal conductivity and thermal diffusivity of Al$_2$O$_3$ nanofluids. *Int. J. Phys. Sci.*, *9*, 514e518.

42. E. Nagy, T. Feczk_o, & B. Koroknai (2007). Enhancement of oxygen mass transfer rate in the presence of nanosized particles. *Chem. Eng. Sci.*, *62*, 7391–7398.

43. J. S. Lee, J. W. Lee, & Y. T. Kang (2015). CO$_2$ absorption/regeneration enhancement in DI water with suspended nanoparticles for energy conversion application. *Appl. Energ*, *143*, 119–129.

44. M. Darabi, M. Rahimi, & A. MolaeiDehkordi (2017). Gas absorption enhancement in hollow fiber membrane contactors using nanofluids: modeling and simulation. *Chemical Engineering and Processing, Process Intensif.*, *119*, 7–15.

6 Numeric Approach of Heat Transfer Application of Nanofluids

6.1 INTRODUCTION

The researchers are working to develop various models for the mathematical justification of the various experimental data available in the literature. Thermo-physical properties are important for the prediction of the behaviour of the heat transfer efficiency of the nanofluids. Researchers have attempted to predict the physical properties like pumping power and viscosity based on nanoparticle concentration and temperature of the system and surroundings [1]. Various attempts have been made by various researchers to develop an efficient model of the viscosity of nanofluids. But, no model so far can predict the wide range of temperature and nanoparticle concentration. Thus, generally, models are developed for the particular application and working conditions only. Multi-layer perceptron-artificial neural network (MLP-ANN) is one such method. Figure 6.1 represents the structure of the MLP-ANN method [2]. For example, Ahmadi et al. [3] used the ANN method for the development of the model for viscosity calculation of silica/water: ethylene glycol nanofluids.

Hybrid nanofluids have been gaining popularity in recent times due to their efficient heat transfer process [4]. The route used to synthesize the hybrid nanofluid is an important parameter of the thermo-physical property of nanofluid. Thus, it becomes complicated to model the nanofluid's performance [5]. Ahmadi et al. [6] attempted to develop the ANN-based model for a hybrid nanofluid of TiO_2-Al_2O_3/water system. Ali et al. highlighted the role of temperature and nanoparticle concentration in the development of numeric models [7].

Various other researchers have attempted to model the ANN-based nanofluid systems [8]. In Chapter 2, we saw the effect of nanoparticle concentration and volumetric flow rate of coolant on the thermal efficiency of flat plate solar collectors. Similarly, the models developed predicted the behaviour. These results are shown in Figure 6.2. Said et al. [9] conducted the thermal efficiency analysis for the linear Fresnel solar collector for the MWCNT/water system and calculated the thermal efficiency of nearly 34% for the 0.3 vol% nanoparticle concentration. As we have discussed in Chapter 2, energy efficiency presents an incomplete picture. The exergy analysis is important to understand the actual potential of the nanofluid system [10]. The wastage of energy occurs due to the increase in the entropy of the

DOI: 10.1201/9781003404767-6

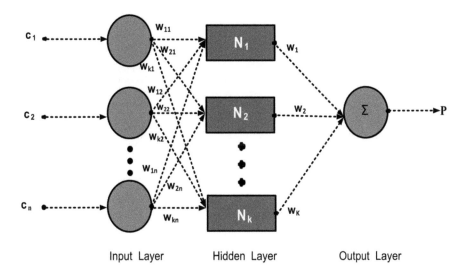

FIGURE 6.1 Structure of multi-layer perceptron-artificial neural network (MLP-ANN) [2].

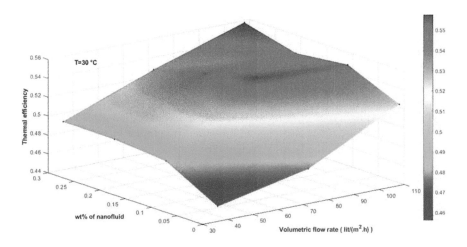

FIGURE 6.2 3D plot of the effect of nanoparticle concentration and volumetric flow rate on thermal efficiency of flat plate solar collector [8].

system. Flow properties and the nanoparticle concentration are important parameters for the increased entropy in the system. Thus, during exergy analysis, we must calculate the nanoparticle concentration and flow properties. These results are shown in Figure 6.3 [11]. Some other parameters like the geometry of the system used, temperature across the system, and energy losses during the process are also important while developing the model for the nanofluid system. These parameters play an important role in the entropy generation within the system [12]. Thus, while developing the model we must consider the exergy destruction and entropy generation rate [13].

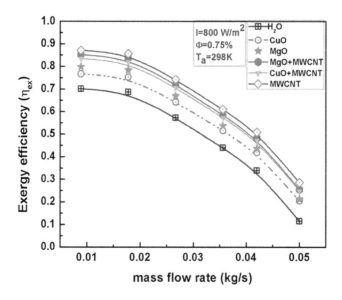

FIGURE 6.3   Exergy destruction rate at (a) 20 MPa pressure and (b) 30 MPa pressure [11].

Thermal devices are used in almost all industrial sectors. These are devices used to achieve total efficiency and productivity. Heat transfer mechanisms can be effective after using these strategies and heat resistance to liquid and air that inhibit heat transfer can be significantly minimized. Integration of types, configurations and geometry of extended surfaces in heat exchangers improves the performance of heat transfer. Other types of cross-sectional pipes like modified pipes such as flat, oval tubes or elliptic tubes also play an important role in the intensification of the heat transfer. Fin and tube heat exchangers (or vehicle heaters) use these tubes to increase the exchange energy and therefore the heat transfer rate. Tourbillon flow devices, such as the longitudinal generators of Delta Wilelet, flexible vertebral generators and rigid vortex generators, tilted longitudinal, a semi-generator of the Delta sheet, delta rectangular vortex generators, prism rectangular vortex generators, Delta fingers and rectangular wings, to be used to create longitudinal vortexes and the angle on heat exchange surfaces. Helical tape inserts and ribs also carry out an important function in breaking the layers of thermal and hydraulic boundary layer and increasing the exchange rate of energy. Nanofluids is one of the methods that can be used for this purpose. The thermal conductivity of the solid particles in the base fluid suspended and the best data transfer properties of the solid phase at mixing. Nanofluids are used at all times of the art, for example in HVAC systems, as refrigerants in electronics, CODUCUBS, the heating of buildings, rooms and defense, medicine, heating bodies cars, nuclear reactor cores, thermal energy deposits, solar thermal conversion (solar, snapshot) and photovoltaic solar thermal systems (PV/T). The use of nanofluids and their disadvantages are the subject of a complete revision search. A revision of demand for condensing nanofluids and evaporation systems was also carried out. In addition to the thermal conductivity, a series of important phenomena such as the movement of the Browny motorcycle, thermophoresis, the geometric parameters of the nanoparticles and the addition of different surface activators play an important role in the supply of the nanofluids with improved thermal-hydraulic performance and remarkable Heat transfer properties. The interface layer responsible for activating the thermal properties of the nanofluids is formed by a complicated interaction between suspended nanoparticles and water molecules. The aforementioned phenomena, in addition to the proper selection of a base fluid that is suited for nanoparticles in terms of kind, size, and shape, have a considerable impact on the nanofluids. Figure 6.4 is the representation of the effect of nanoparticle concentration and mass flow rate on the

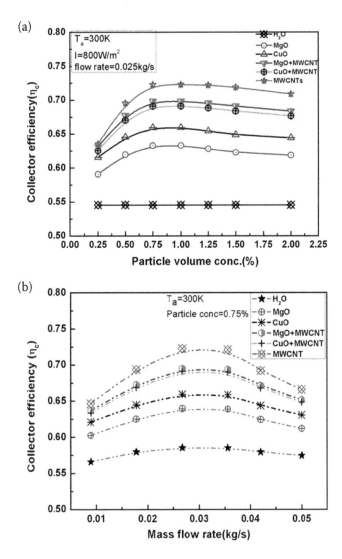

FIGURE 6.4 Effect of (a) nanoparticle concentration and (b) mass flow rate on collector efficiency [11].

collector efficiency. The preparation technique, which might be one or two steps, has a significant impact on the thermophysical properties. During dispersion, agglomeration of nanoparticles and sedimentation in the liquid must be prevented; otherwise, the thermo-physical properties of nanofluids will be altered.

To maintain the properties of nanofluids transport, an adequate analysis of stability and measurement should be used. The particle load has a significant effect on thermal conductivity, specific thermal capacity, the expansion coefficient and viscosity. The optimal concentration of nanoparticles has a significant effect on these qualities. The excessive particle charge accelerates the kinetics of agglomeration and sedimentation, increases nanofluids viscosity and causes a particulate bond on heat transfer surfaces. As a result, the properties of heat transfer are altered and thermal power is lower than that of the conventional liquid. In addition, fouling

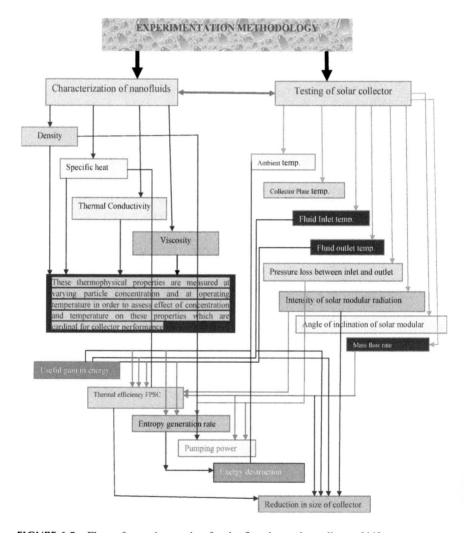

FIGURE 6.5 Flow of experimentation for the flat plate solar collector [11].

includes the behaviour and viscosity of the highest work fluid ever greater pressure loss and therefore more pumping power. Figure 6.5 represents the general flow of experimentation for the nanofluids application in the flat plate solar collector.

6.2 NUMERIC EVALUATION OF SOLAR PANEL PERFORMANCE

Solar panels are designed by the exergy analysis of the system. Exergy is the amount of useful energy. This analysis is important for the calculation of the efficiency of solar panels. Following are some important strategies for the calculation of energy and exergy of solar panel systems.

6.2.1 IMPORTANT FORMULAS FOR ENERGY ANALYSIS

The thermal efficiency of the solar panel is the ratio of energy utilized by the solar panel to the total solar energy incident on the solar panel, i.e.,

$$\eta = \frac{Q_u}{A_c I_T} \tag{6.1}$$

Here, η is the thermal efficiency of the solar panel, Q_u is energy utilized by solar panels (W), A_c is the total area of the solar panel (m^2) and I_T is incident radiation (W/m^2).

Generally, energy utilized can be calculated by simple energy balance as given below,

$$Q_u = mCp(T_{out} - T_{in}) \tag{6.2}$$

Here, m is the mass flow rate of nanofluids, Cp is the specific heat of nanofluids and T_{out} and T_{in} are the outlet and inlet temperature of nanofluids respectively. Thus, Equation (6.2) can be written as

$$\eta = \frac{mCp(T_{out} - T_{in})}{A_c I_T} \tag{6.3}$$

The energy utilized can be calculated by the following formula for energy calculation of solar panels.

$$Q_u = A_c F_R [I_T(\tau\alpha) - U_L(T_i - T_a) \tag{6.4}$$

Here, F_R is friction factor and $\tau\alpha$ is product of transmittance-absorbance parameter. Comparing these equations, we can get the following equation.

$$\text{Thus,} \quad \eta = \frac{mCp(T_{out,f} - T_{in,f})}{A_c I_T} = F_R(\tau\alpha) - U_L F_R\left(\frac{T_i - T_a}{I_T}\right) \tag{6.5}$$

From the above equation, we can easily compare the performance of various nanofluids in the solar panels. The nature of these lines is linear. We need to perform experimentations to calculate the outlet temperature at different conditions.

These outlet values at different conditions will be required to calculate the efficiency of nanofluids in the solar panels. When the efficiency values are plotted on the y-axis and $\left(\frac{T_i - T_a}{I_r}\right)$ values on the x-axis, the intersection of the lines at the y-axis is the value of $F_R(\tau\alpha)$. The negative sign in the equation represents the decreasing values of efficiency with an increase in the $\left(\frac{T_i - T_a}{I_r}\right)$ value. The pressure drop along the length of the collector is shown in the following equation:

$$\Delta p = f \cdot \frac{\rho \cdot V^2 \Delta l}{2d} \tag{6.6}$$

Here, f is a friction factor, V is the velocity of the liquid, ρ is the density of the liquid, L is the length of the channel, and d is the diameter of channels used for the solar panels.

Here,

$$f = \frac{64}{Re} \tag{6.7}$$

Re is Reynolds number. The pumping power required for the operation is shown in the following equation. The general relation between the particle concentration and pumping power is shown in Figure 6.6.

$$\text{Pumping Power} = \left(\frac{m}{\rho_{nf}}\right) \times \Delta p \tag{6.8}$$

As we know, though, nanofluids increase the thermal performance of the nanofluids, but, the use of nanofluids also leads to an increase in the pumping power required. Thus, performance evaluation criterion (PEC) is used to compare the performance

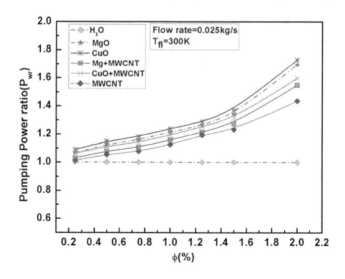

FIGURE 6.6   Effect of nanoparticle concentration on the pumping power [11].

of nanofluids. The thermal performance is evaluated by the Nusselt number and the pumping power is evaluated by the friction factor [14].

$$PEC = \frac{\left(\frac{Nu_{nf}}{Nu_{bf}}\right)}{\left(\frac{f_{nf}}{f_{bf}}\right)^{\frac{1}{3}}} \tag{6.9}$$

6.2.2   DATA REDUCTION FOR EXERGY ANALYSIS

Exergy destruction is calculated by applying the exergy balance over the system. The following equation shows the exergy balance of the system

$$\sum Ex_{in} - \sum Ex_{out} = \sum Ex_{dest} \tag{6.10}$$

Ex_{in} is total exergy entering the system, Ex_{out} is total exergy leaving the system and Ex_{dest} is total exergy lost during the process. Exergy enters the system along with the heat absorbed from the sun ($Ex_{in,\ rad}$) and exergy available with the incoming working fluid ($Ex_{in,f}$). Exergy leaves the system with an outlet working fluid stream ($Ex_{out,f}$). Destruction of exergy occurs at the absorber plate due to the difference in the temperature of the sun and the absorber plate ($Ex_{dest,s-p}$), exergy loss due to heat lost from the plate to the surroundings ($Ex_{dest,\ lost}$) and exergy also lost due to differences in the temperature of the plate and working fluid ($Ex_{dest,p-f}$).

The following equation is used to calculate the exergy lost during the process [15]

$$Ex_{out,f} - Ex_{in,f} = mC_p \left[(T_{out,f} - T_{in,f}) - T_a \ln\left(\frac{T_{out,f}}{T_{in,f}}\right) \right] \tag{6.11}$$

The following equation is used to calculate the exergy inlet from the sun

$$Ex_{in,rad} = I_T A_P \left[1 - \left(\frac{T_a}{T_S}\right) \right] \tag{6.12}$$

The following equation is used to determine the exergy efficiency of the solar panels.

$$\eta_{ex} = \frac{mC_p \left[(T_{out,f} - T_{in,f}) - T_a \ln\left(\frac{T_{out,f}}{T_{in,f}}\right) \right]}{I_T A_P \left[1 - \left(\frac{T_a}{T_S}\right) \right]} \tag{6.13}$$

The following equation is also used to determine the exergy of nanofluid-based solar panels. The general trends of exergy efficiency are shown in Figure 6.7.

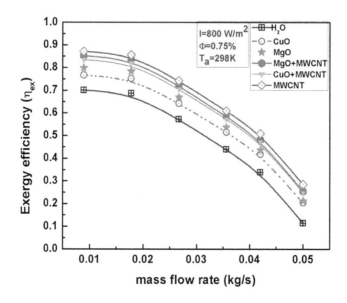

FIGURE 6.7   Exergy efficiency [11].

$$\eta_{ex} = \frac{mCp\left\{\left[(T_{in,f} - T_a - S/U_L)\left(\exp\left(\frac{U_L A_P F'}{mC_P}\right) - 1\right)\right] - T_a \ln\left(\frac{(T_{in,f} - T_a - S/U_L)\left(\exp\left(\frac{U_L A_P F'}{mC_P}\right) - 1\right)}{T_{in,f}} + 1\right)\right\}}{I_T A_P\left[1 - \left(\frac{T_a}{T_S}\right)\right]}$$ (6.14)

Exergy absorbed by the absorber plate is given as

$$Ex_{abs} = I_T A_P \eta_o \left[1 - \left(\frac{T_o}{T_P}\right)\right]$$ (6.15)

Exergy destruction is calculated by subtracting the inlet exergy from the outlet exergy

$$Ex_{dest,s-p} = I_T A_P \left[1 - \eta_o + \eta_o \left(\frac{T_o}{T_P}\right) - \left(\frac{T_o}{T_P}\right)\right]$$ (6.16)

The loss of exergy in the surroundings is calculated by the following equation.

$$Ex_{dest,lost} = U_l A_P (T_P - T_o) \left[1 - \left(\frac{T_o}{T_P}\right)\right]$$ (6.17)

Exergy transferred from the plate to the working fluid is calculated by the following equation.

$$Ex_{p-f} = mC_p(T_{out,f} - T_{in,f})\left[1 - \left(\frac{T_o}{T_P}\right)\right]$$ (6.18)

Exergy destructed during the absorber plate to working fluid is calculated by the following equation.

$$Ex_{dest,p-f} = mC_p(T_{out,f} - T_{in,f})\left[1 - \left(\frac{T_o}{T_P}\right)\right]$$ (6.19)

The exergy can be related to entropy generation by following the equation [16]

$$Ex_{dest} = T_a S_{gen}$$ (6.20)

where

$$S_{gen} = mC_{p,nf} T_a \ln\frac{T_{f,out}}{T_{f,in}} - \frac{mT_a\Delta p}{T_{f,in}\rho}$$
$$- [1 - (\tau\alpha)]\left\{\frac{I_T A_c}{T_a}\left[1 - \frac{4}{3}\left(\frac{T_a}{T_s}\right) + \frac{1}{3}\left(\frac{T_a}{T_s}\right)^4\right]\right\}$$ (6.21)

The general trend for the entropy growth rate for the mass flow rate is represented in Figure 6.8.

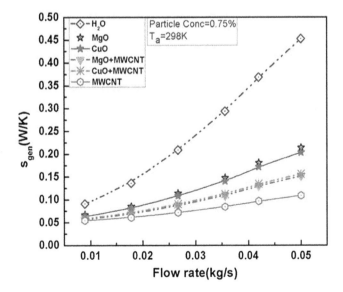

FIGURE 6.8 Entropy growth rate for flow rate [11].

The following equation can also be used to calculate the exergy efficiency [17]

$$\eta_{ex} = 1 - \frac{T_a S_{gen}}{\left(1 - \frac{T_a}{T_{sur}}\right) Q_s} \tag{6.22}$$

6.2.3 MATHEMATICAL MODELS

Afzal and Aziz [18] numerically investigated unsteady magneto-hydrodynamic boundary layer slip flow and heat transfer of nanofluids in solar collectors. They assumed thermal conductivity as a function of temperature. To reduce the governing boundary value problem to a system of nonlinear ordinary differential equations they used the similarity transformation technique. The various results obtained from this technique are shown in Figure 6.9.

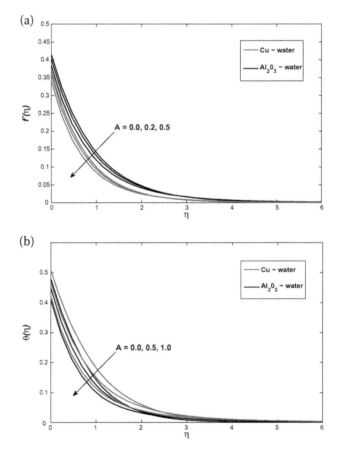

FIGURE 6.9 (a) Velocity profile for alumina/water and Cu/water nanofluids. (b) Temperature profile for alumina/water and Cu/water nanofluids [18].

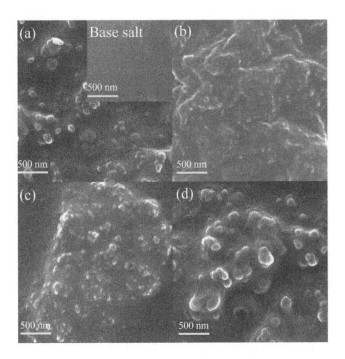

FIGURE 6.10   SEM images of (a) salt + 0.5% silica, (b) salt + 1% silica, (c) salt + 1.5% silica and (d) salt + 2% silica [19].

Hu [19] proposed a freeze-drying method to formulate solar nanofluids. The SEM images of nanoparticles used in this study are shown in Figure 6.10. To simulate the forced convective heat, transfer modified lattice Boltzmann model was used. They got 1 weight% mass fraction as optimal. The result shows that the Shah-London correlation can be used for the design of solar heat exchangers [20,21].

Two types of models are developed from the generalized equations using the time fractional form namely, the Caputo-Fabrizio model and the Atangana-Baleanu derivative. By both models, it can be concluded that the average increase in the Nusselt number is 4–5%. Alumina-based nanofluids have shown better Nusselt numbers at different time parameters. Similarly, the behaviour of both models is different at different time parameter values.

6.3   NUMERIC EVALUATION OF CAR RADIATOR

The computational fluid dynamics (CFD) study of the car radiator was conducted by Bai et al. [22]. They compared the performance of nanofluids with the base fluid. According to their study, they have suggested that copper-based nanofluids can deliver better results than any other nanofluids. As the nanoparticle concentration is increased in the simulations, the performance of the car radiator also increases significantly. The temperature of the coolant is reduced by 45%, after the increase of 5% nanoparticle concentration. But, the workload on the pump increases by 6%. Huminic et al. [23] used a commercial CFD tool of ANSYS to evaluate the nanofluid's performance in the car radiator. They have reported that the nanofluids are better than the base fluid in terms of convective heat transfer rate. The velocity of coolant, the temperature of the system and nanoparticle loading are important parameters for the heat transfer. Vajjha et al. [24] used different software tools like Gambit to analyze the flow behaviour and heat transfer of nanofluids. The alumina and copper-based nanofluids with water and ethylene glycol as base fluid with 60:40 composition had an increase in the heat transfer coefficient by 77% and pumping power was reduced by 70% for equivalent performance. Figure 6.11 represents the variation of the Nusselt number and heat transfer coefficient for the Reynolds number.

Ray et al. [25] used MATLAB for the mathematical modeling of results obtained using the ε-NTU method for different nanofluids of alumina, copper oxide and silica nanoparticles suspended in the water and ethylene glycol mixture in 60:40 proportion. The result showed that ideally the pumping power can be reduced by 35%, 33% and 26%, respectively, the surface area also increases by 7.4%, 7.2% and 5.2%. Park et al. [26] conducted a simulation study in U-tube radiators; the technique used for this purpose is known as a SIMPLE technique for irregular geometry. This technique is mainly used for the calculation of flow behavior and temperature effect over the flow.

Bharadwaj et al. [27] conducted the simulations using the ANSYS FLUENT software. The inlet temperature is kept at 50 °C and the inlet velocity is set at 1.5 m/s. The author used graphene oxide nanoparticles in water and ethylene glycol. The experimental data required for this study is taken from the various literature available. This data was useful for the prediction of the co-relations among various fluid

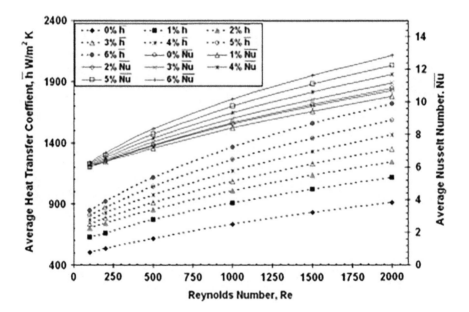

FIGURE 6.11 Effect of change of Reynolds number on Nusselt number and heat transfer coefficient [24].

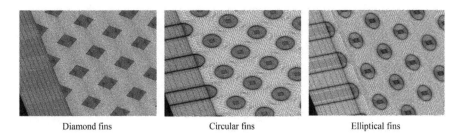

Diamond fins   Circular fins   Elliptical fins

FIGURE 6.12   Various geometries used by Ambreen et al. [28].

properties. The grid independence tests were also important to ensure the minimum simulation time would be taken by software. The effectiveness of the radiator is increased by using more nanoparticle concentration. Ambreen et al. [28] conducted the simulation study with the alumina and copper oxide-based nanofluids of different morphology. The shape of the particle was circular, elliptical and rhombus. Figure 6.12 represents the different geometries used in this study.

Naiman et al. [29] conducted a simulation study for the nanocellulose alumina nanoparticle in the base fluid of water and ethylene glycol. The one-dimensional simulation study is conducted by the author. The role of nanofluid velocity is highlighted in this study. An increase in the flow rate increases the heat transfer efficiency. These results were also validated by the experimental results. The nanofluids synthesized for the experimentations are prepared using a two-step method. Benedict et al. [30] studied the doped nanoparticles of alumina and TiO_2 in the mixture of ethylene glycol and water with a 60:40 ratio for the car radiator application. The doped nanoparticle is added with plant-based carbon nanocellulose to synthesize the hybrid nanofluids. An increase in thermal conductivity is observed in the mono-nanoparticle-based nanofluids and hybrid nanofluids. But, the relative performance of hybrid nanofluids is good. The maximum increase in thermal conductivity is observed in the case of alumina/CNC-based nanofluids. An increase of 108% is reported in the literature. Sahoo et al. [31] conducted a mathematical study using a wavy-finned radiator. Three different types of nanofluids are used for this study. Copper oxide of thermal conductivity of 40 W/m·K graphene of thermal conductivity of 5000 W/m·K and carbon nanotubes of thermal conductivity of 3000 W/m·K. The different shapes of nanoparticles are also used for the study. Among all these nanofluids hybrid nanofluids of carbon nanotube doped on graphene and copper oxide had the best performance. The results were good at a dilute concentration of nanoparticles. Sahoo et al. [32] also conducted the study of different nanoparticles like copper oxide, silver, copper, ferric oxide, and titanium dioxide. These nanoparticles were added with the alumina in equal proportion. And results showed that nanofluids with the alumina and silver nanoparticles have shown better performance compared to the other nanofluid system. The recent advances and other progress are shown in Table 6.1.

From the studies, it is evident that the performance of the radiator is dependent on the various parameters as shown in Figure 6.13.

TABLE 6.1

Some Co-relations Developed for the Nanofluid-Based Car Radiator System [33]

Sr. No.	Nanofluid System	Co-relation	Relation Valid for	Reference
1	Al$_2$O$_3$/ethylene glycol+water (50:50)	$Nu = 0.70091(1 + 10.4576\emptyset^{0.9}Pe^{0.353548})Re^{0.495596}Pr^{0.06752}$	valid for flat tube geometry; forced convection and laminar flow	[34]
2	Al$_2$O$_3$/ethylene glycol+water (40:60) CuO/water+ethylene glycol (40:60)	$Nu_{avg} = 1.953\left(Re.\,Pr\frac{D_h}{Z}\right)^{\frac{1}{3}}$ For $\left(Re.\,Pr\frac{D_h}{Z}\right) \geq 33.33$ $Nu_{avg} = 4.364 + 0.0722\left(Re.\,Pr\frac{D_h}{Z}\right)$ For $\left(Re.\,Pr\frac{D_h}{Z}\right) < 33.33$	valid for flat tube geometry; forced convection and entrance region	[24]
3	Al$_2$O$_3$/ethylene glycol+water (60:40) CuO/water+ethylene glycol (60:40)	$Nu_{nf} = 0.023Re_{Dh}^{0.8}Pr^{0.3}(1 + 0.1771\emptyset^{0.1466})$ For $1.988 \leq Pr \leq 13.44$ $3000 \leq Re \leq 8000$ $0 \leq \emptyset \leq 0.06$	valid for flat-tube geometry; forced convection and fully developed turbulent flow	[35]

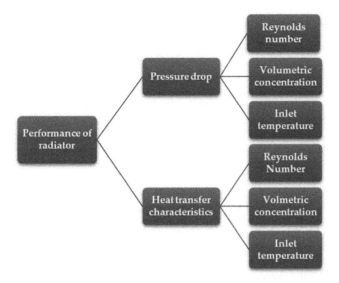

FIGURE 6.13   Important parameters affecting the car radiator performance [33].

Apart from the solar panels and car radiator, we have seen the application of nanofluids in the boiling process in Chapter 4. The general discussion on the numeric approach to the boiling application of nanofluids is briefly given in Chapter 4.

6.4   CONCLUSION

In this chapter, numerical approaches for the various heat transfer applications of nanofluids are discussed in detail. The parameters affecting the experimental results are discussed with the various examples of heat transfer operations of nanofluids. The application of nanofluids in car radiators and solar panels is discussed in detail. The solar panel and car radiator are examples of heat exchange operations. Various co-relations governing these applications are reported in this chapter. Important formulas for the calculation of exergy and energy efficiency of the solar panels are given in detail.

REFERENCES

1. H. Babar, M. U. Sajid, & H. M. Ali (2019). Viscosity of hybrid nanofluids: A critical review. *Thermal Science*, *23*(3 Part B), 1713–1754.
2. M. Ramezanizadeh, M. H. Ahmadi, M. A. Nazari, M. Sadeghzadeh, & L. Chen (2019). A review on the utilized machine learning approaches for modeling the dynamic viscosity of nanofluids. *Renewable and Sustainable Energy Reviews*, *114*, 109345
3. M. H. Ahmadi, M. Sadeghzadeh, H. Maddah, A. Solouk, R. Kumar, & K. W. Chau (2019). Precise smart model for estimating dynamic viscosity of SiO$_2$/ethylene glycol–water nanofluid. *Engineering Applications of Computational Fluid Mechanics*, *13*(1), 1095–1105.
4. M. H. Ahmadi, M. Ghazvini, M. Sadeghzadeh, M. A. Nazari, & M. Ghalandari (2019). Utilization of hybrid nanofluids in solar energy applications: A review. *Nano-Structures & Nano-Objects*, *20*, 100386.

5. M. Gupta, V. Singh, S. Kumar, S. Kumar, Dilbaghi. N., & Z. Said (2018). Up to date review on the synthesis and thermophysical properties of hybrid nanofluids. *Journal of Cleaner Production*, *190*, 169–192.

6. M. Sadeghzadeh, H. Maddah, M. H. Ahmadi, A. Khadang, M. Ghazvini, A. Mosavi, & N. Nabipour (2020). Prediction of thermo-physical properties of TiO_2-Al_2O_3/water nanoparticles by using artificial neural network. *Nanomaterials*, *10*(4), 697.

7. M. U. Sajid, & H. M. Ali (2018). Thermal conductivity of hybrid nanofluids: A critical review. *International Journal of Heat and Mass Transfer*, *126*, 211–234.

8. M. Sadeghzadeh, M. H. Ahmadi, M. Kahani, H. Sakhaeinia, H. Chaji, & L. Chen (2019). Smart modeling by using artificial intelligent techniques on thermal performance of flat-plate solar collector using nanofluid. *Energy Science & Engineering*, *7*(5), 1649–1658.

9. M. Ghodbane, Z. Said, A. A. Hachicha, & B. Boumeddane (2019). Performance assessment of linear Fresnel solar reflector using MWCNTs/DW nanofluids. *Renewable Energy*, *151*(May 2020), 43–56.

10. M. Abid, M. S. Khan, & T. A. Hussain Ratlamwala (2019). Thermodynamic performance evaluation of a solar parabolic dish assisted multigeneration system. *Journal of Solar Energy Engineering*, *141*(6), 1–10.

11. S. K. Verma, A. K. Tiwari, S. Tiwari, & D. S. Chauhan (2018). Performance analysis of hybrid nanofluids in flat plate solar collector as an advanced working fluid. *Solar Energy*, *167*, 231–241.

12. M. Abid, M. S. Khan, & T. A. H. Ratlamwala (2020). Comparative energy, exergy and exergo-economic analysis of solar driven supercritical carbon dioxide power and hydrogen generation cycle. *International Journal of Hydrogen Energy*, *45*(9), 5653–5667.

13. N. Akram, R. Sadri, S. N. Kazi, M. N. M. Zubir, M. Ridha, W. Ahmed, & M. Arzpeyma (2019). A comprehensive review on nanofluid operated solar flat plate collectors. *Journal of Thermal Analysis and Calorimetry*, *139*(2020), 1309–1343.

14. M. Ghodbane, Z. Said, A. A. Hachicha, & B. Boumeddane (2019). Performance assessment of linear Fresnel solar reflector using MWCNTs/DW nanofluids. *Renewable Energy*, *151*(May 2020), 43–56.

15. F. Jafarkazemi & E. Ahmadifard (2013). Energetic and energetic evaluation of flat plate solar collectors. *Renewable Energy*, *56*, 55–63.

16. Z. Said, M. A. Alim, & I. Janajreh (2015). Exergy efficiency analysis of a flat plate solar collector using graphene based nanofluid. In *IOP conference series: Materials science and engineering* (Vol. *92*, No. 1, p. 012015). IOP Publishing.

17. A. Bejan (2002). Fundamentals of exergy analysis, entropy generation minimization, and the generation of flow architecture. *International Journal of Energy Research*, *26*(7), 545–565.

18. K. Afzal & A. Aziz (2016). Transport and heat transfer of time dependent MHD slip flow of nanofluids in solar collectors with variable thermal conductivity and thermal radiation. 10.1016/j.rinp.2016.09.017

19. Y. Hu, Y. He, H. Gao, & Z. Zhang (2019). Forced convective heat transfer characteristics of solar salt-based SiO_2 nanofluids in solar energy applications. 10.1016/j.applthermaleng.2019.04.109

20. K. Khanafer & K. Vafai (2011). A critical synthesis of thermophysical characteristics of nanofluids. 10.1016/j.ijheatmasstransfer.2011.04.048

21. N. A. Sheikh, F. Ali, I. Khan, M. Gohar, & M. Saqib (2018). On the applications of nanofluids to enhance the performance of solar collectors: A comparative analysis of Atangana-Baleanu and Caputo-Fabrizio fractional models. DOI 10.1140/epjp/i2017-11809-9

22. M. Bai, Z. Xu, & J. Lv (2008). *Application of nanofluids in engine cooling system* (No. 2008-01-1821). SAE Technical Paper.
23. G. Huminic & A. Huminic (2012). (No. 2012-01-1045). SAE Technical Paper. The cooling performances evaluation of nanofluids in a compact heat exchanger.
24. R. S. Vajjha, D. K. Das, & P. K. Namburu (2010). Numerical study of fluid dynamic and heat transfer performance of Al_2O_3 and CuO nanofluids in the flat tubes of a radiator. *International Journal of Heat and Fluid Flow, 31*(4), 613–621.
25. D. R. Ray & D. K. Das (2014). Superior performance of nanofluids in an automotive radiator. *Journal of Thermal Science and Engineering Applications, 6*(4), 041002.
26. K. W. Park & H. Y. Pak (2002). Flow and heat transfer characteristics in flat tubes of a radiator. *Numerical Heat Transfer: Part A: Applications, 41*(1), 19–40.
27. B. R. Bharadwaj, K. S. Mogeraya, D. M. Manjunath, B. R. Ponangi, K. R. Prasad, & V. Krishna (2018, April). CFD analysis of heat transfer performance of graphene based hybrid nanofluid in radiators. In *IOP Conference Series: Materials Science and Engineering* (Vol. 346, No. 1, p. 012084). IOP Publishing.
28. T. Ambreen, A. Saleem, H. M. Ali, S. A. Shehzad, & C. W. Park (2019). Performance analysis of hybrid nanofluid in a heat sink equipped with sharp and streamlined micro pin-fins. *Powder Technology, 355*, 552–563.
29. I. Naiman, D. Ramasamy, & K. Kadirgama (2019). Experimental and one dimensional investigation on nanocellulose and aluminium oxide hybrid nanofluid as a new coolant for radiator. In *IOP Conference Series: Materials Science and Engineering* (Vol. 469, No. 1, p. 012096). IOP Publishing.
30. F. Benedict, A. Kumar, K. Kadirgama, H. A. Mohammed, D. Ramasamy, M. Samykano, & R. Saidur (2020). Thermal performance of hybrid-inspired coolant for radiator application. *Nanomaterials, 10*(6), 1100.
31. R. R. Sahoo (2021). Effect of various shape and nanoparticle concentration based ternary hybrid nanofluid coolant on the thermal performance for automotive radiator. *Heat and Mass Transfer, 57*(5), 873–887.
32. R. R. Sahoo, P. Ghosh, & J. Sarkar (2017). Performance analysis of a louvered fin automotive radiator using hybrid nanofluid as coolant. *Heat Transfer—Asian Research, 46*(7), 978–995.
33. F. Abbas, H. M. Ali, T. R. Shah, H. Babar, M. M. Janjua, U. Sajjad, & M. Amer (2020). Nanofluid: Potential evaluation in automotive radiator. *Journal of Molecular Liquids, 297*, 112014.
34. S. K. Saripella, W. Yu, J. L. Routbort, & D. M. France (2007). *Effects of nanofluid coolant in a class 8 truck engine* (No. 2007-01-2141). SAE Technical Paper.
35. R. S. Vajjha, D. K. Das, & D. R. Ray (2015). Development of new correlations for the Nusselt number and the friction factor under turbulent flow of nanofluids in flat tubes. *International Journal of Heat and Mass Transfer, 80*, 353–367.

7 Numeric Approach of the Mass Transfer Application of Nanofluids

7.1 INTRODUCTION

Global warming has become a major concern for humans and the earth. The atmospheric CO_2 concentration has grown from 289 parts per million in the 17th century to 406 parts per million in 2017, as a result of human activity and the industrial revolution. This has caused an increase by $20\,°C$. Climate warming during the 21st century is foreseen to grow further as a result of greenhouse gas emissions. CO_2 is the primary greenhouse gas emitted through industries and by humans. Compared to the other principal heat-trapping gases produced as a result of human activity, CO_2 stays in the atmosphere for longer. Methane (CH_4) emissions take 10 years to depart the atmosphere (to be converted to CO_2), but nitrous oxide emissions take a century (N_2O). Due to the increase in the usage of natural resources such as fossil fuels by humans, there is a remarkable increase in the carbon dioxide concentration and it keeps on increasing year by year and warms the planet. To reduce the global warming and to overcome the environmental issues, CO_2 has to be removed. There are various methods for CO_2 sequestration, some of which are cryogenic methods, membrane separation, adsorption, absorption, biological fixation and wet scrubbing. Figure 7.1 represents the various methods available for carbon dioxide capture and separation. Absorption, adsorption and membranes are the most widely used techniques for carbon dioxide separation. The absorption of CO_2 is divided into two methods, the first is chemical absorption and the second is physical absorption. The use of amines and ammonia are examples of chemical absorption of CO_2 and methods like the selexol process and rectisol process are examples of the physical absorption process. Adsorption is also an important method used for CO_2 separation. Beds of zeolite, alumina and other materials are used for the adsorption. The regeneration of CO_2 is done by the pressure swing or temperature swing methods. Apart from absorption and adsorption, the use of the membranes is also done for the CO_2 separations.

The use of chemical looping, cryogenics, membranes, etc. has not yet reached a level where they are feasible. Figure 7.2 represents the conventional CO_2 absorption technique at the industrial level. Adsorption and absorption techniques are now the most popular strategies to remove CO_2. However, adsorption requires too much energy for the desorption process, making it inappropriate for use in large-scale

 DOI: 10.1201/9781003404767-7

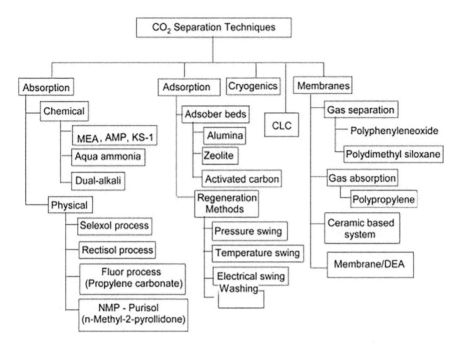

FIGURE 7.1   Various methods available for the CO_2 separation [1].

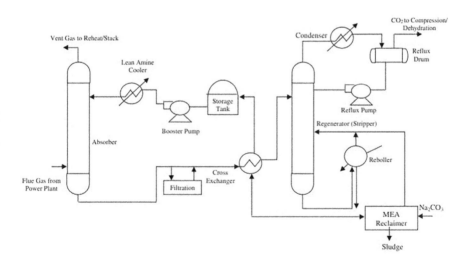

FIGURE 7.2 CO_2 absorption process used on the industrial scale [1].

systems. Each method has its advantages and disadvantages. Some methods are extremely energy consuming, have high cost and corrosive and thus it is essential to make progress with new techniques with a high efficiency of CO_2 capture and also the cost should be less. Among these various techniques, due to the little energy needed by gas-liquid, there has been a significant advancement in CO_2

absorption by gas-liquid. According to the type of absorbent absorption, methods can be divided into two types: the chemical type and other is physical absorption. When there is no chemical reaction between the solute and the solvent, physical absorption takes place. While in the chemical absorption impurities are removed from the gas phase and dissolved into liquid. It is possible to considerably reduce the costs of the absorption process by using CO_2 absorption enhancement methods.

A new approach called nanotechnology is being used extensively in many different energy systems. In the past few years, the importance of the revolutionary method of CO_2 absorption or conversion has increased by employing nanofluids. The solvent contains tiny nanoparticles that have been dissolved in it. These particles have a significantly larger surface area than typical particles do. Because more gas and liquid molecules may react at once due to an improvement in surface area, the rate of mass transport rises.

Various characteristics of nanofluids such as nanoparticle size, concentration, type and base fluid affect the efficiency of absorption and changing these parameters increases or decreases efficiency. Studies on gas absorption use nanofluids, and to improve gas absorption, an external magnetic field is also created in the absorption column.

The enhancement can also be improved by showing up of magnetic field with nanofluids. Magneto-hydrodynamic nanofluids or ferro-fluids are a sloughy-like mixture of magnetic nanoparticles such as Fe_3O_4 in a base fluid. By using an external magnetic field, the mass transfer properties can be improved. When the magnetic Fe_3O_4 particles are dissolved in the base solvent and subjected to an external magnetic field, separation into nanoparticles occurs from the nanofluids. As a result, the surface area of interaction increases due to which mass transfer increases. There is an increase in the random motion of the nanoparticles resulting in a fast mass transfer rate in the presence of the magnetic field.

The mass transfer enhancement can also be greatly improved by using nanofluids in the presence of an external electric field. According to the experimental conclusions, the external electric field can greatly increase the CO_2 absorption rate in dilute nanofluids, while it weakens mass transfer in concentrated nanofluids.

7.2 MECHANISM OF NANOFLUIDS IN MASS TRANSFER ENHANCEMENT OF ABSORPTION

The primary process for improving mass transfer in nanofluids is the grazing effect which is also called as shuttle effect and is described as the absorption and desorption of gas molecules on the surface of nanoparticles from the interface of gas-liquid to the bulk fluid. The decreased bubble size in nanofluids revealed that nanoparticles are the primary factor for improving mass transport through enhanced area of contact between the two phases, mass diffusivity, and mass transfer coefficient. They also micro-convection as one of the key elements accelerating the rate of mass transport. Three types of mechanisms can be inferred for the enhancement of mass transfer. Various mechanisms of CO_2 absorption are represented in Figure 7.3.

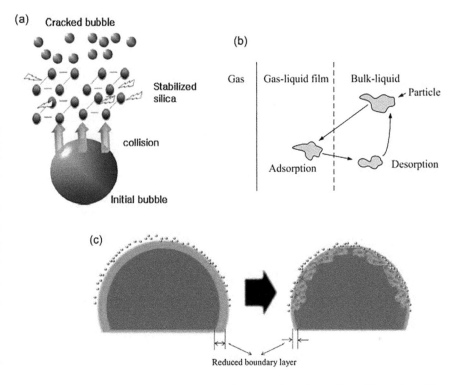

FIGURE 7.3 The mechanism of CO_2 absorption by (a) bubble breaking, (b) grazing effect and (c) hydro-dynamic effect [2].

7.2.1 BUBBLE BREAKING

These effects are present in the surface of contact of the gas-liquid phase. A very thin diffusion layer is formed due to the nanoparticle's involvement with the interface of the liquid-gas layer. This results in the gas diffusion into the layer of fluid and blending with the bulk of liquid, the mass transfer coefficient is then increased. The gas bubble is smaller when nanoparticles are present.

7.2.2 GRAZING EFFECT

As per the two-film theory, gas-liquid interface films regulate the movement of material between two phases. The diffusion through the liquid film is controlled by the gradient of concentration between the solutes present on both sides of the liquid film whereas the diffusion through the gas film is related to the gradient of concentration among the solutes present on both sides of the gas film.

7.2.3 HYDRO-DYNAMIC EFFECT

Using particle adsorption and desorption, more gas is transferred from the liquid bulk to the gas phase. Adsorption takes place in the diffusion layer of liquid and gas during this phase, and desorption happens in the volume of liquid, increasing the mass transfer coefficient. These effects are explained by the bubble dynamics in Chapter 5 in detail.

As we have discussed in Section 5.3 (Chapter 5), the increase in the nanoparticle concentration leads to an increase in carbon dioxide absorption. But generally, after 0.1 vol% of nanoparticle concentration, the increase in CO_2 absorption efficiency becomes steady. This is because the excess presence of nanoparticles decreases the nanofluid's ability to absorb the carbon dioxide. These results are shown in Figure 5.13 (Chapter 5). The figure represents the absorption efficiency of the three solvents and their nanofluids. The trend of absorption is same for all three solvents. All the concentrations of the di-ethanol amine-based nanofluids have more efficiency than the water-based nanofluids and NaOH-based nanofluids.

Thus, we can conclude that the stability of nanofluids is important for the better performance of the nanofluids. From the comparison of Figures 5.12 and 5.13 (Chapter 5),we can say that because of the increased stability of nanofluids, absorption of carbon dioxide is greater in di-ethanol amine-based nanofluids. As we have discussed in Section 5.3 (Chapter 5), the role of nanoparticle size is not very important.

7.2.4 FLOW RATE STUDY

The gas flow rate is an important parameter for the nanofluid's performance. As we have discussed and seen in Figure 5.7 (Chapter 5), the increase in the gas flow rate is not recommended. But, those results are for the membrane-based geometry. From Figure 5.13 (Chapter 5), we can see that, an increase in the gas flow rate results in better mass transfer. This is because, in the bubble column, the higher the turbulence, the more mass transfer is observed. Increased gas velocity creates turbulence in the system. In Figure 5.13, section (a) is carbon dioxide absorbed by the water-based nanofluids with different concentrations of fly ash. Section (b) is CO_2 absorption values for the caustic soda and section (c) is absorption values for the di-ethanol amine. From all these results, we can conclude that the increase in the velocity of gas increased the absorption efficiency.

7.2.5 REGENERATION STUDY

The regeneration of nanofluids is done by the thermal method. Solar energy is used to regenerate the carbon dioxide from the nanofluids. Nanofluids are allowed to pass through the microchannels in the solar panels. As temperature increases entrapped carbon dioxide gets released into the air. This study is conducted at various nanofluid velocities.

The results obtained for the desorption studies are shown in Figure 5.14 (Chapter 5) as water-based nanofluids (a), caustic soda-based nanofluids (b) and DEA-based nanofluids (c). From the results, it is evident that the turbulence of the nanofluids generated at the high liquid flow is an important parameter for the increased mass transfer. The desorption in the DEA-based nanofluids is higher because amines are comparatively easily degraded at high temperatures. The presence of nanoparticles increases the storage of thermal energy.

7.2.6 BUBBLE DYNAMICS

The bubble dynamics during the absorption of carbon dioxide is a very important parameter. The smaller size of the bubble ensures a higher surface area for mass transfer. We have used a high-speed camera to capture the bubbles generated during the experimentation. For a better understanding of nanofluids performance. The parameters like bubble diameter, and bubble frequency are measured to understand the mechanism in detail. Bubble dynamics are dependent on the column structure, packing material, solvent and gas flow rates. This data is useful for the development of the various mathematical models. The images were processed by ImageJ software for the determination of the bubble diameter. Images represented in Figure 7.4 are real-time images of the carbon dioxide absorption.

7.2.6.1 Bubble Departure Diameter

From the images, the bubble diameter is measured, and the results of these studies are shown in Figure 5.16 (Chapter 5). The mechanism of the enhancement of carbon dioxide absorption can be explained by the results obtained from the bubble dynamics study. The grazing effect and hydrodynamic effect on the boundary layer discussed in

(a)   (b)

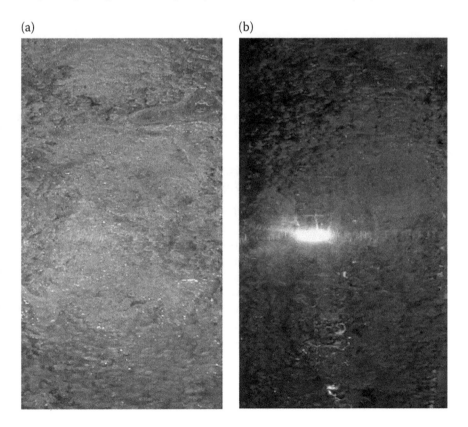

FIGURE 7.4 Real-time images captured during the experimentations: (a) water-based nanofluids; (b) DEA-based nanofluids.

Section 5.2 (Chapter 5) can be explained by these results. As we can see, bubble diameter decreases with the increase in the nanoparticle concentration in the nanofluids. The trend of carbon dioxide is the same in water-based nanofluids and amine-based nanofluids.

7.2.6.2   Cycle Times

The cycle time of the bubble is measured at a very low gas flow rate. The flow rate is maintained in such a way that the bubble should reach the top of the column. This time is recorded on camera and the timings manually measured. For the current study, the gas flow rate is maintained at the flow rate of 1×10^{-5} m³/sec. The total cycle is represented in Figure 5.17 (Chapter 5). So we can see that nanofluids play a very important role in the breaking of the bubbles and movement of the bubble. The Brownian motion of nanoparticles is the main reason behind the decreased cycle time required for this bubble movement. The results are similar in the water-based nanofluids and amine-based nanofluids. Amine-based nanofluids have comparatively less cycle time.

7.2.6.3   Frequency of Bubble (*f*)

The frequency of the bubble is calculated from the following equation.

$$f = \frac{1}{T_c} \qquad (7.1)$$

Here, T_c is bubble cycle time. The frequency of the bubble can be measured by this analysis. The flow rate of gas is again kept constant and bubble cycle time is measured to evaluate the bubble departure frequency. The bubble departure frequency is comparatively higher in the amine-based nanofluids.

7.3   ADVANTAGES OF USING NANOFLUIDS FOR CO₂ ABSORPTION OVER CONVENTIONAL METHODS

There are several traditional techniques for CO_2 absorption or separation, some of which have limitations, such as vortex and venturi contactors' traditional spray columns and packed columns. Take the packed column method into consideration; it has problems with flooding and entrainment. They can only be used with low-viscosity absorbents because, more crucially, they have serious issues with viscose absorbents. The ineffective surface-to-volume ratio and back mixing of gas of conventional spray towers result in a reduction in the effectiveness of absorption. In liquid and gas phases, vortex and venturi contactors have a high surface area of contact, but they have high operating and fundamental costs.

7.4   MATHEMATICAL APPROACHES FOR THE NANOFLUIDS APPLICATION IN CO₂ ABSORPTION STUDY

7.4.1   MODEL DEVELOPMENT

Various attempts have been made by researchers for the mathematical model development of CO_2 absorption study. The hollow fiber membrane based study is discussed in this section. The use of hollow fiber-based membranes has increased in recent years for the nanofluids-based absorption system. The basic geometry of a hollow fiber-based membrane is divided into three parts. First is the tube side of the hollow fiber membrane, this side is also called as liquid side of the membrane. The second part is the shell side of the membrane. This part is also called as gas side of the membrane and the third part is the membrane. In this part the porous membrane is present. The stepwise mass transfer of carbon dioxide also occurs in these three sections. In the first step of mass transfer, the gas gets diffused at the outer layer of the membrane. Then in the second step, this gas gets diffused to the surface of the membrane. Then, in the third step of mass transfer, this gas gets diffused to the tube side or the bulk of the liquid. In this third step of mass transfer, all the respective reactions will take place. The detailed geometry of the hollow fiber membrane is represented in Figure 7.5. The cross-sectional area of this membrane contactor is shown in Figure 7.6. For simplicity, in the mathematical calculations, the radius of the tube is termed $r1$, the radius of the membrane is termed $r2$ and the radius of the shell will be termed as $r3$. This geometry is shown in Figure 7.7. The shell radius is calculated by Happel's model [3].

$$r_3 = \left(\frac{1}{\frac{\pi r_2}{H^2}} \right)^{0.5} r_2 \tag{7.2}$$

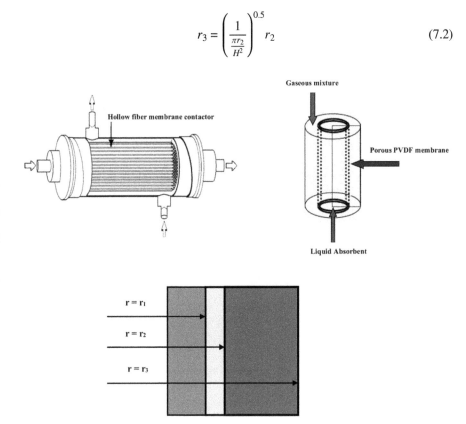

FIGURE 7.5   A schematic representation of membrane contactor [4].

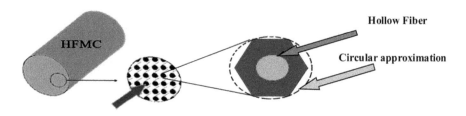

FIGURE 7.6 Cross-sectional area of hollow fiber membrane [4].

The radius of the module is represented by H.
 The assumptions used for the modeling are as follows:

1. The simulations are conducted for constant temperature and steady-state conditions.
2. The membrane is considered as a wet.
3. For gas-liquid equilibrium, Henry's law is used.
4. For the prediction of liquid velocity, Happel's model is used.
5. The gas used is assumed to be ideal gas.
6. The flow condition is assumed to be a fully developed laminar flow condition.
7. The gas-liquid is assumed to be in the counter-current flow arrangement.

The same geometry with the same assumptions is used for the desorption study also. Only, the gas concentration reduces and the nanofluids are passed through the tubes.

7.4.2 GAS SIDE EQUATIONS

The following equation is used for the mass transfer from the gas side to the liquid side [5]:

$$\frac{\partial C_{CO_2}}{\partial t} = -\nabla N_{CO_2} + R_{CO_2} \tag{7.3}$$

In this equation, C is concentration, N is mass transfer flux and R is the reaction rate.

$$D_{CO_2,\ gas} \left[\frac{\partial^2 C_{CO_{2,gas}}}{\partial r^2} + \frac{1}{r}\frac{\partial C_{CO_{2gas}}}{\partial r} + \frac{\partial^2 C_{CO_{2gas}}}{\partial z^2} \right] = V_{z,gas}\frac{\partial C_{CO_{2gas}}}{\partial z} \tag{7.4}$$

Here, the velocity term is solved by following equations [6]

$$\rho(V_{z-shell}.\nabla)V_{z-shell} = -\nabla.\ [-pI + \mu(\nabla V_{z-shell} + (\nabla V_{z-shell})^T)] \tag{7.5}$$

$$\nabla V_{z-shell} = 0 \tag{7.6}$$

7.4.3 MEMBRANE SIDE EQUATIONS

The following equation is used as a continuity equation of the membrane side [7]

$$D_{CO_2,m}\left[\frac{\partial^2 C_{CO_2,m}}{\partial r^2} + \frac{1}{r}\frac{\partial C_{CO_2,m}}{\partial r} + \frac{\partial^2 C_{CO_2,m}}{\partial z^2}\right] = 0 \qquad (7.7)$$

The partial weight assumed for the above equation is represented by the following figure.

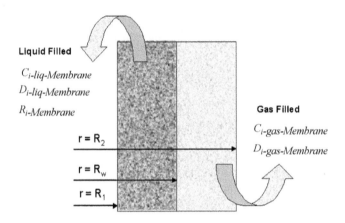

Liquid Filled
$C_{i\text{-}liq\text{-}Membrane}$
$D_{i\text{-}liq\text{-}Membrane}$
$R_{i\text{-}Membrane}$

$r = R_2$
$r = R_w$
$r = R_1$

Gas Filled
$C_{i\text{-}gas\text{-}Membrane}$
$D_{i\text{-}gas\text{-}Membrane}$

FIGURE 7.7 Membrane assumed for the numeric model [7].

7.4.4 LIQUID EQUATIONS

The following equation is used as a continuity equation for the liquid side:

$$D_{CO_2,T}\left[\frac{\partial^2 C_{CO_2,T}}{\partial r^2} + \frac{1}{r}\frac{\partial C_{CO_2,T}}{\partial r} + \frac{\partial^2 C_{CO_2,T}}{\partial z^2}\right] = V_{z,T}\frac{\partial C_{CO_2,T}}{\partial z} - R_{CO_2} + \frac{k_p \alpha_p (C_{CO_2,t} - C_s)}{1 - \varphi}$$

$$(7.8)$$

As nanofluids are used as a solvent here, it is important to exclude the Brownian motion from the equation for desorption study [8].

$$D_{n,f} = D_{b,f}(1 + 640Re^{1.7}Sc^{1/3}\varphi) \qquad (7.9)$$

φ is the concentration of nanoparticles in the nanofluids. The following equation is used to calculate the Reynolds number and Schmidt number [9].

$$Re = \left(\frac{18KT\rho^2}{\pi d_p \rho_p \mu}\right)^{0.5} \qquad (7.10)$$

$$Sc = \frac{\mu}{\rho D} \tag{7.11}$$

Boltzmann's constant is represented by K.

The grazing effect phenomena are calculated by the following equation [10].

$$\varphi \rho_p V_z \frac{\partial q}{\partial z} = k_p \alpha_p \left(C_{CO_{2,t}} - C_s \right) \tag{7.12}$$

In this equation, q is the number of gas bubbles adsorbed on the surfaces of the nanoparticles. Thus, by using a suitable adsorption isotherm like Langmuir isotherm, we can calculate the amount of carbon dioxide adhered to the nanoparticle surface.

$$q = q_m \frac{k_d C_s}{1 + k_d C_s} \tag{7.13}$$

7.4.5 RESULTS OBTAINED FROM THE MODEL

During the numeric analysis of the above nanofluids, the mass transfer is assumed as a convective mass transfer. Boundary conditions are set as per the requirement of the geometry. The thermodynamic equilibrium is assumed at the gas-liquid interphase. A no-slip condition is assumed at the membrane wall. For the numeric analysis, COMSOL 5.4 software is used. A grid independence test is performed to estimate the grid size. The solver used for this analysis is the PARDISO solver.

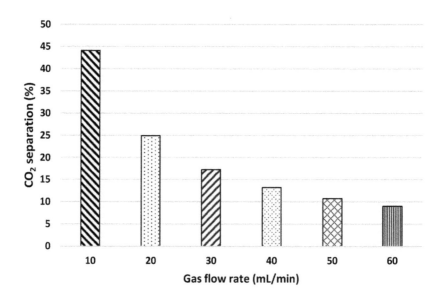

FIGURE 7.8 Impact of gas flow rate on carbon dioxide absorption [11].

7.4.5.1   Impact of Gas Flow Rate and Liquid Flow Rate on Absorption and Desorption of Carbon Dioxide

The gas flow rate should be kept minimum, the high flow rate doesn't give the required contact time for the mass transfer. Thus, the increased gas flow decreases the CO_2 absorption. These results are shown in Figure 7.7 [11]. As we know, the longer the contact time, the more surface area will be available for the mass transfer. Thus,

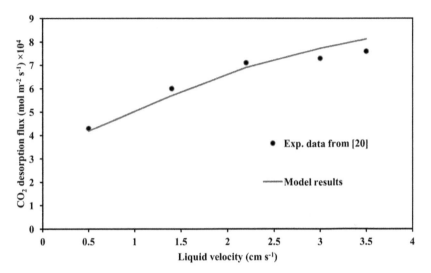

FIGURE 7.9   Impact of liquid flow rate on carbon dioxide desorption [12].

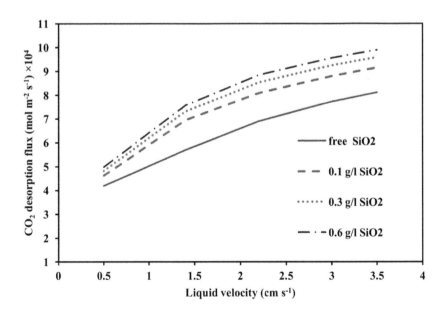

FIGURE 7.10   Effect of liquid velocity on CO_2 desorption [12].

more CO_2 will be separated from the gas side. Similarly, in the desorption study, an increase in the liquid flow rate is not recommended beyond the optimum limit. At a very low liquid flow rate, desorption flux will be also less. This is represented by Figures 7.8 and 7.9 [12]. The addition of nanoparticles has a positive impact on the absorption and desorption study. These results are reported in Figure 7.10.

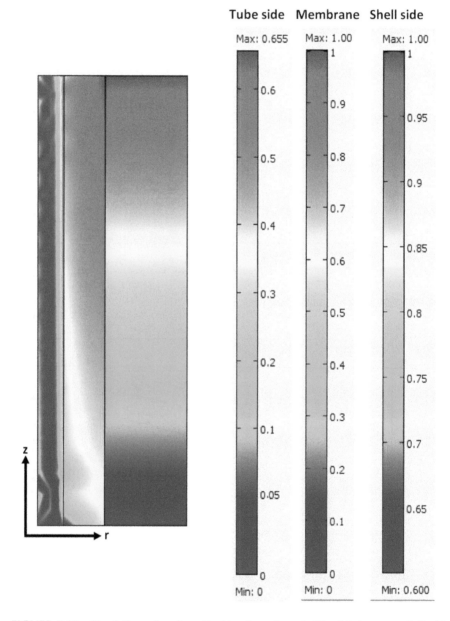

FIGURE 7.11 Simulation of carbon dioxide absorption at 10 mL/min gas and liquid velocity [11].

7.4.5.2 Impact of Concentration of Solvent and Nanoparticles

The more the nanoparticle and solvent concentration present in the nanofluids, the more mass transfer is observed [13,14]. The simulation results are shown in Figure 7.11. From this result, we can see that, the concentration decreases very rapidly along the radius. Similar results are observed in the desorption operation also. The results of desorption studies are reported in Figure 7.12 [15–17]. The approach used for the CO_2 absorption and desorption study can also be used for the

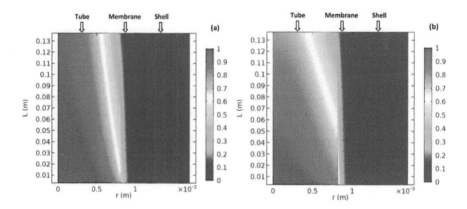

FIGURE 7.12 Simulation results of carbon dioxide desorption at (a) without solvent and (b) 0.6 g/L concentration of silica and 3 kmol/m^3 MEA concentration [12].

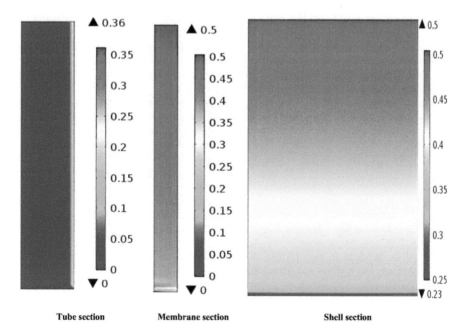

Tube section   Membrane section   Shell section

FIGURE 7.13 CO_2 concentration in silica/water nanofluids [17].

extraction operations in the membrane contactor. The length of the membrane is also a very important parameter of the efficiency of extraction. Generally, the mass transfer increases with the increase in membrane length [6]. The results of mass transfer in radial direction are seen in Figure 7.13. The maximum CO_2 concentration is t the inlet of the shell, then it starts to decrease [17].

REFERENCES

1. A. A. Olajire (2010). CO_2 capture and separation technologies for end-of-pipe applications–A review. *Energy*, *35*, 2610–2628.
2. J. S. Lee, J. W. Lee, & Y. T. Kang (2015). CO_2 absorption/regeneration enhancement in DI water with suspended nanoparticles for energy conversion application. *Applied Energy*, *143*, 119e129.
3. J. Happel (1959). Viscous flow relative to arrays of cylinders. *AIChE Journal*, *5*(2), 174–177.
4. A. T. Nakhjiri, A. Heydarinasab, O. Bakhtiari, & T. Mohammadi (2018). Modeling and simulation of CO_2 separation from CO_2/CH_4 gaseous mixture using potassium glycinate, potassium argininate and sodium hydroxide liquid absorbents in the hollow fiber membrane contactor. *Journal of Environmental Chemical Engineering*, *6*, 1500–1511.
5. S. Eslami, S. M. Mousavi, S. Danesh, & H. Banazadeh (2011). Modeling and simulation of CO_2 removal from power plant flue gas by PG solution in a hollow fiber membrane contactor. *Advances in Engineering Software*, *42*, 612–620.
6. M. Ghadiri, A. Hemmati, A. T. Nakhjiri, & S. Shirazian (2020). Modelling tyramine extraction from wastewater using a non-dispersive solvent extraction process. *Environmental Science and Pollution Research*, *27*, 39068–39076.
7. R. Faiz & M. Al-Marzouqi (2009). Mathematical modeling for the simultaneous absorption of CO_2 and H_2S using MEA in hollow fiber membrane contactors. *Journal of Membrane Science*, *342*, 269–278.
8. E. Nagy, T. Feczkó, & B. Koroknai (2007). Enhancement of oxygen mass transfer rate in the presence of nanosized particles. *Chemical Engineering Science*, *62*, 7391–7398.
9. R. Prasher, P. Bhattacharya, & P. E. Phelan (2005). Thermal conductivity of nanoscale colloidal solutions (nanofluids). *Physical Review Letters*, *94*, 025901.
10. L. Sumin, X. Min, S. Yan, & D. Xiangjun (2013). Experimental and theoretical studies of CO_2 absorption enhancement by nano-Al_2O_3 and carbon nanotube particles. *Chinese Journal of Chemical Engineering*, *21*, 983–990.
11. Y. Cao, S. M. S. Alizadeh, M. T. Fouladvand, A. Khan, A. T. Nakhjiri, Z. Heidari, … & A. B. Albadarin (2021). Mathematical modeling and numerical simulation of CO_2 capture using MDEA-based nanofluids in nanostructure membranes. *Process Safety and Environmental Protection*, *148*, 1377–1385.
12. H. N. Mohammed, S. M. Ahmed, H. Al-Naseri, & M. Al-Dahhan (2021). Enhancement of CO_2 desorption from MEA-based nanofluids in membrane contactor: Simulation study. *Chemical Engineering and Processing-Process Intensification*, *168*, 108582.
13. V. S. Sefidi & P. Luis (2019). Advanced amino acid-based technologies for CO_2 capture: A review. *Industrial & Engineering Chemistry Research*, *58*, 20181–20194.
14. T. N. G. Borhani, A. Azarpour, V. Akbaria, S. R. W. Alwi, & Z. A. Manan (2015). CO_2 capture with potassium carbonate solutions: A state-of-the-art review. *International Journal of Greenhouse Gas Control*, *41*, 142–162.

15. P. Zare, P. Keshavarz, & D. Mowla (2019). Membrane absorption coupling process for CO_2 capture: application of water-based ZnO, TiO2, and multi-walled carbon nanotube nanofluids. *Energy Fuels*, *33*, 1392–1403.
16. M. Rezakazemi, M. Darabi, E. Soroush, & M. Mesbah (2019). CO_2 absorption enhancement by water-based nanofluids of CNT and SiO_2 using hollow-fiber membrane contactor. *Separation and Purification Technology*, *210*, 920–926.
17. A. Marjani, A. T. Nakhjiri, A. S. Taleghani, & S. Shirazian (2020). Mass transfer modeling absorption using nanofluids in porous polymeric membranes. *Journal of Molecular Liquids*, *318*, 114115.

8 Progress and Challenges to Nanofluids' Future Prospects

8.1 INTRODUCTION

Unstable nanofluids lead to the inefficient performance of heat transfer and mass transfer operations. Stable nanofluids offer better thermal properties. Agglomerated nanoparticles in unstable nanofluids are nothing but resistant to heat transfer operation and also create an extra burden on the pumping and transportation of fluids on an industrial scale. Researchers have studied various parameters and methods to improve the stability of nanofluids. However, there is a lot of scope for the research and detailed mechanism for the stable nanofluids. But, still, the nanofluids do not stay in the suspension for a longer time.

Some researchers like Sunder et al. [1] and Hussein [2] have developed nanofluids with stability for 60 days. Similarly, Yarmand et al. [3] developed stable hybrid nanofluids of graphene nanoparticle–silver/water with a 60-day stability period. But, for more application of nanofluids in industrial scale, this stability period should be increased. New strategies for nanofluids need to be explored. The author has developed a nanocomposite for better stability. Figure 8.1 represents the strategy to synthesize the novel nanocomposite material. Figure 8.2 represents the TEM images of synthesized nanocomposite material.

Akhgar and Toghraie [4] used the sedimentation/photograph technique to evaluate the stability of mono nanoparticle-based nanofluids of TiO_2 and MWCNT nanoparticles. The author reported that at a pH value of 9, the TiO_2 nanofluids showed stability for 48 hours. However, MWCNT-based nanofluids didn't show stability for 48 hours. This is due to the hydrophobic nature of MWCNT. Thus, to improve the stability author has added cetrimonium bromide (CTAB) surfactant in the nanofluids. To manage the stability Asadi et al. [5] used the zeta potential method for the stability measurement of $Mg(OH)_2$-MWCNT/engine oil nanofluids. This is because, for dark-colourednanofluids, the use of the sedimentation/photograph method or centrifugation method is not practical.

The strategies for improving stability and measuring it have been documented in the literature [6]. Wei et al. [7] developed stable nanofluids of $SiC-TiO_2$/oil nanofluids of 52 mV zeta potential and they remained stable for 10 days. The results of the stability of nanofluids are shown in Figure 8.3. Similarly, Safi et al. [8] developed the hybrid nanofluids of MWNT and TiO_2 with a 47 mV zeta potential. Wei et al. [9], reviewed the parameters affecting the stability of nanofluids and

DOI: 10.1201/9781003404767-8

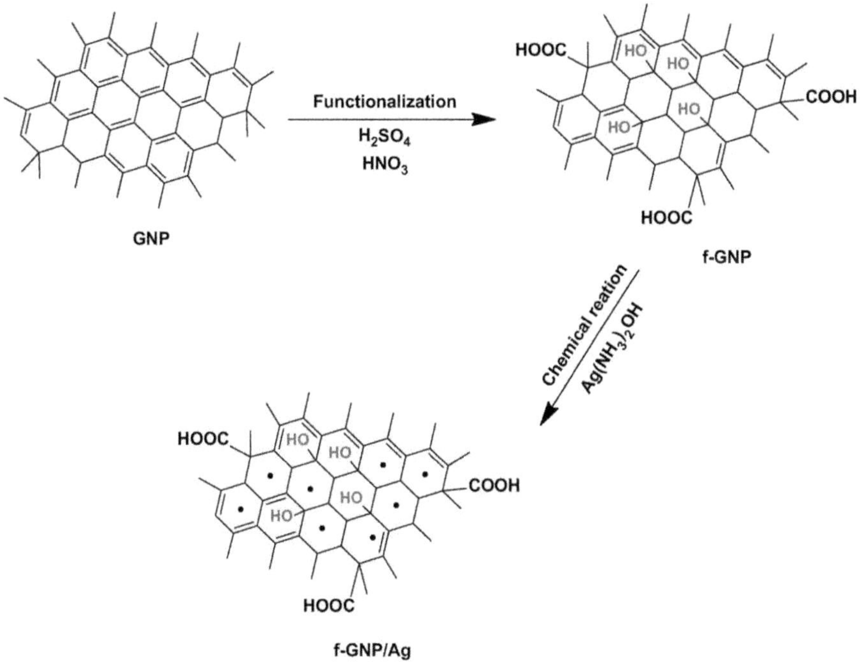

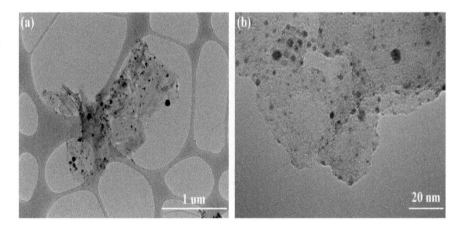

FIGURE 8.1 Synthesis route of graphene/silver nanocomposite material [3].

FIGURE 8.2 TEM images of synthesized nanocomposite material [3].

noted that it is important to develop stable nanofluids first. Otherwise, there is no point in using the nanofluids without stability for a longer time. The results obtained for the SiC-TiO$_2$/oil nanofluids are shown in Figure 8.4. From the graphs, we can see that the stability of nanofluids decreases over time.

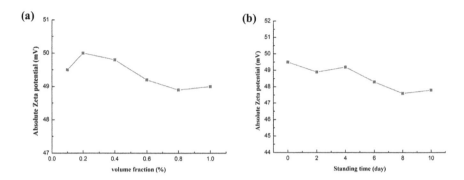

FIGURE 8.3   Zeta potential value for the (a) nanoparticle concentration and (b) time [7].

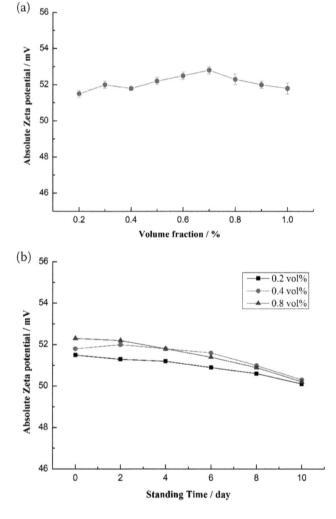

FIGURE 8.4   Zeta potential values for time and nanoparticle concentration for the SiC-TiO$_2$/oil nanofluids [9].

8.2 APPLICATION OF NANOFLUIDS

Because of the improved heat transfer and mass transfer efficiency, nanofluids have found application in various industrial processes. Following are a few examples of nanofluids applications.

8.2.1 TRANSPORTATION

The use of nanofluids for the thermal management systems of automobiles has been implemented by various researchers. Various studies have proved that nanofluids have shown remarkable enhancement in heat transfer performance. This nanofluids-based system reduces the size of conventional automobile engine cooling systems. This reduced size of the cooling systems is important for fuel economy. Ethylene glycol is added to the water as a conventional coolant. Various researchers have published data for the nanofluids application specifically for the automobile industry. Koçak et al. [10] used copper and silver-doped TiO_2-based nanofluids and compared the obtained results. The author reported that the doping mechanism and nanoparticles used for doping are important parameters for the nanofluids study. Their results showed that in the case of silver-doped TiO_2-based nanofluids have the best results compared to other nanofluids. But, the copper-doped TiO_2 nanofluids decreased the performance lower than TiO_2-based nanofluids. Similar studies of the nanofluids for car radiators with other nanoparticles were also conducted. For example, Peyghambarzadeh et al. [11], studied the Al_2O_3/water nanofluids for radiator fluids; the geometries used by the author for the study are shown in Figure 8.5. The experimental and numerical studies have reported an enhancement in heat transfer. Olivera et al. [12], used MWCNT/water nanofluids and other nanofluids for this study.

8.2.2 ELECTRONIC APPLICATIONS

The heat released from electronic devices is managed by two methods. The first method is to design the optimum size of the cooling system and the other method is to use the high heat transfer medium for heat release. For the first method, researchers developed the micro-channel-based heat sink for electronic cooling systems. Nanofluids are an example of the second method. Nanofluids offer a better heat transfer rate than conventional heat transfer mediums. Various researchers have shown the various nanofluid systems. Jang et al. [13] used diamond and copper

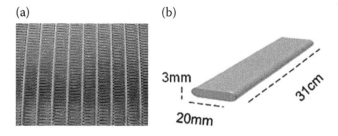

(a)   (b)

3mm

20mm

31cm

FIGURE 8.5   (a) Fins and tubes of car radiators used in the car. (b) Tube geometry used for the study [11].

(a)

(b)

(c)

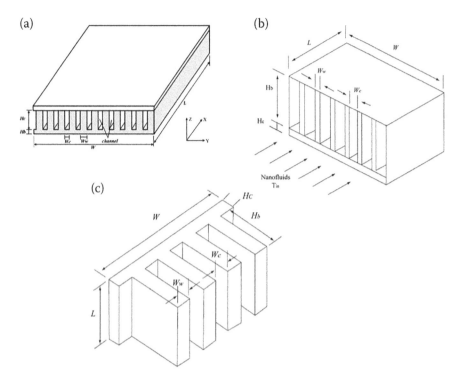

FIGURE 8.6 Mini-channel-based heat sink used for the nanofluids application for electronics cooling [14].

nanoparticles in water-based nanofluids for thermal management and better thermal performance is observed than water as a coolant in the system.

The electronics sector is rapidly growing every year. Thus, it is very important to develop efficient cooling systems. The micro-channel based cooling systems are already being used in electronic cooling systems. The use of nanofluids has also increased in recent years. Nanofluids are the best fit for the micro-channel-based heat release. Ijam et al. [14] performed various experiments with the SiO_2 and TiO_2-based nanofluids as a coolant in electronic devices. Various other nanoparticles can also be used for electronic cooling applications. The mini-channel-based heat sink used for this study is shown in Figure 8.6. Maganti et al. [15] proposed the hybrid nanofluid system for this purpose. Ag/FeO$_3$-water and Al$_2$O$_3$/water are some of the examples for this purpose.

8.2.3 DEFENSE AND SPACE

Space and defense sectors require efficient cooling systems to work in compact spaces in military equipment and satellites. Aircraft require the best thermal management systems with high efficiency that are lightweight. The scope of the nanofluids in this application is very high. These equipment need cooling systems that can handle high heat flux of magnitude of 10 MW/m². In various research papers, it has been proven that nanofluids can offer better performance for pool boiling and flow boiling of nanofluids. Nanofluids can also be used for the thermal management of military vehicles, high-energy weaponry and submarines.

8.2.4 INDUSTRIAL COOLING APPLICATIONS

Industrial equipment used in various industrial operations also need cooling systems. Mechanical energy is also released in these machinery in the form of heat. Prolonged heat release may result in machine damage. Thus, nanofluids-based systems can effectively manage the thermal performance of nanofluids. Generally, the water is reused for this purpose in the industry. The temperature of the machinery is controlled by the water available in the industry .

Nanofluids are also important in the lubrication of machinery applications. This application is also called as tribological application of nanofluids. Nanofluids form a protective film in tribological applications. This property of nanofluids is mainly used in the petroleum industry for enhanced oil recovery. As the size of nanoparticles is very small, they are not rigid and do not easily erode the mechanical parts. Nanofluids have less friction factor than other tribological solvents. Thus, the nanofluids-based systems are used for the enhanced oil recovery application. Generally, ferro-fluids are used for this purpose. Nanofluids with Fe_3O_4 nanoparticles are called ferro-fluids. These fluids have magnetic properties. Nanofluids can also be used as a working fluid in the drilling operation at energy reservoirs. The temperature during such operation may increase to thousands of degrees Celsius. Chahta et al. [16] developed a drilling process

FIGURE 8.7 Experimental setup required for the application of nanofluids as a drilling fluid [16].

using nanofluids as a working fluid. These nanofluids are referred to as nanofluids minimal quality lubricant (NFMQL) and drilling operations have better operation with the nanofluids-based working fluid. The torque required to conduct the drilling operation and required thrust is better with the nanofluids. This experimentation setup is shown in Figure 8.7.

8.2.5 ENERGY APPLICATIONS

Nanofluids can be effectively used in energy absorption and storage applications like solar panels. The energy storage capability of nanofluids has been identified by various researchers. A lot of research is going on to a develop more efficient energy capture and storage medium using nanofluids. Energy storage is an important aspect of these studies. The development of more efficient solar panels and more efficient working fluid is the objective of various researchers working in this area.

Nanofluids are also used to decrease the smoke during combustion operations. Nanofluids also increase the heat of combustion in some cases [17]. Nanofluids are also used in the polymer exchange membrane as a coolant [18]. The setup of the polymer-exchange membrane is shown in Figure 8.8

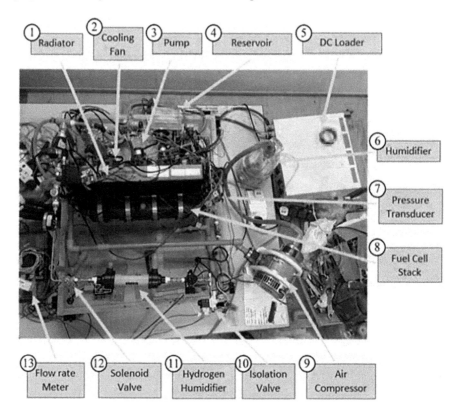

FIGURE 8.8 Schematic diagram and experimental setup used for the energy application of nanofluids [18].

8.2.6 BIOMEDICAL APPLICATIONS

Nanofluids are also important in biomedical operations. Various types of anti-microbial nanoparticles are synthesized by the one-step method route and used as anti-microbial compounds for biomedical applications. These nanoparticles are coated with other suitable compounds as per the requirement. Various studies have shown that silver nanoparticle has shown great application as a coating agent for the ZnO nanoparticle. These coated nanoparticles are primarily used for anti-microbial activity. Various other modifications were also reported in the literature for this application. Bacterial membranes are also used to improve the performance of these nanoparticles. The nanofluids made from these nanoparticles are synthesized in one step and are useful to prevent bacterial growth. Gold nanoparticles are also used for biomedical applications. Along with this gold nanoparticles are useful for controlled drug release. Similarly, carbon nanotubes and graphene nanoparticles are used for these purposes.

8.3 CHALLENGES

As we have discussed in the chapter, the main challenge to the large-scale use of nanofluids is the stability of nanofluids over longer periods. Before using nanofluids, one must know the specific application conditions as nanofluids performance is heavily dependent on the operational parameters. Alteration in operational parameters leads to a decrease in the performance of the nanofluids. Thermophoresis is one of the challenges for using nanofluids. Nanofluids are mainly used for thermal applications. Thermophoresis leads to a rapid decrease in the nanofluid's stability. Thus, a careful analysis of temperature gradient and pressure drop needs to be conducted to achieve maximum heat performance. Synthesis of the nanofluids should be done carefully, otherwise, the contamination during synthesis or application of nanofluids may lead to ineffective thermal performance of the nanofluids.

In the literature, it is observed that thermal conductivity has received more attention than other thermo-physical properties. However, the synergetic effect of all the properties results in the enhanced thermal efficiency of the system. Apart from this, there are other factors that need to be addressed for the application of nanofluids. The experimental results and numeric results do not agree with each other in many cases. The stability of nanofluids for a longer period is a major concern for the researchers. It is important to research to resolve these issues soon. This is because it has been already more than 25 years since nanofluids were first touted for its potential but these issues are still unresolved.

Stability of the nanofluids is considered as the most important challenge. The synthesis methods of nanofluids play vital roles in the stability of nanofluids. In Chapter 1, we have seen that the pH and temperature of nanofluids are crucial for the better stability of nanofluids. The carefully engineered suspension may prevent the early agglomeration of nanoparticles in the nanofluids. The analysis of nanofluids needs some time. Thus, the nanofluids with shorter periods of stability are difficult to analyze [19]. Thus, more efforts are needed to develop a highly effective nanofluid system. The attractive forces between the nanoparticles need to be nullified to synthesize the better suspension of nanofluids [20].

Pressure drop across the coolant length is an important parameter for the application of nanofluids [21]. The pressure drop and pumping power are dependent on each other. Viscosity and density are important physical properties of nanofluids that affect the pumping power and pressure drop of the system. Thus, it is important to note that, the increase in physical properties like viscosity and density will lead to an increase in pressure drop [22]. This is one of the challenges of the application of nanofluids in any system. Erosion and corrosion due to the nanofluids are also important challenges for the nanofluids. Erosion due to metal oxide-based nanoparticles may occur in the prolonged use of nanofluids. Erosion of the system is dangerous because it not only hampers the overall performance of the system but may permanently damage the system. These challenges weaken the candidacy of nanofluids as the best replacement for conventional coolants and solvents for heat and mass transfer. Following are the important challenges that need to be overcome for the large-scale application of the nanofluids.

8.3.1 Uncertainty in Characterization Results Reported in the Literature

The uncertainty in the reported results leads to a false basis for the experimentation. For example, in the characterization of nanofluids, generally, the SEM, TEM and XRD are used for the characterization of nanoparticles. But, most of the time, the focus of the beam is set in such a way that the results do not show the exact picture of all nanoparticles or nanofluids region. While preparing the sample for TEM analysis, the sample is allowed to dry on the disc. This leads to the agglomeration of nanoparticles and thus inaccurate results are obtained. Various studies have reported DLS analysis but the details of pH value and refractive index value used during the analysis are not reported in most of the literature. There are certain limitations of DLS analysis, as it gives the results for low concentrations of nanoparticles. It is thus, difficult to estimate the stability of relatively high concentrations of nanoparticles in nanofluids. Sharifpur et al. [23] used various methods to estimate the stability of nanofluids. They have used the zeta potential method, viscosity method, UV spectroscopy method and sedimentation method. These results are shown in Figure 8.9.

8.3.2 Failure of Models to Estimate Thermo-Physical Properties

Various researchers have attempted to model the nanofluid systems based on the data reported in the literature. However, no model is efficient enough to predict the thermo-physical properties of nanofluids for a wide range of parameters. A slight change in the property leads to deviation from the model. The change in the viscosity and thermal conductivity are more prone to the operating conditions [24,25]. Thus, it is important to understand that, the reported values of viscosity and thermal conductivity may differ from actual values. The physical properties like density and specific heat are based on the law of mixtures, thus, they are easy to model.

The method used for the measurement of thermal conductivity i.e., the hot wire method is very dynamic. The results may vary for operating conditions [26]. This

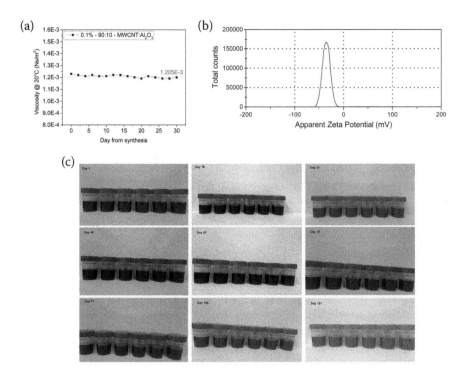

FIGURE 8.9 Various methods for evaluation of nanofluids stability: (a) viscosity measurement method, (b) zeta potential analysis and (c) sedimentation method [23].

inability of the mathematical models leads to confusion regarding the effectiveness of designed nanofluids [27]. The datasheets of the nanofluids are not available in the literature. It is necessary to develop such datasets for the various nanofluid systems [28].

8.3.3 Uncertainty of Results Reported in the Literature

Various research papers in the literature reported contradictory results in many cases. Some papers even conducted the studies at limited parameters and the performance of the nanofluids is reported on these incomplete parameters. These results lead to confusion in mathematical modeling. Because of the nanoparticles in the base fluids, the viscosity of nanofluids increases significantly. This increase in the viscosity results in a decrease in pressure drop and an increase in the pumping power [29]. Thus, the results in the literature must have to be verified before developing any models or assumptions for further studies.

8.3.4 Unrealistic Basis of Reported Studies

In the literature, while conducting the numeric analysis, researchers generally, assume the nanofluids as a homogeneous mixture. This basis of the study eliminates the actual mechanism of the nanofluids and different results are seen. Thus, it becomes difficult to use these results for actual results obtained from the experimental studies. Nowadays, publishers publish research articles related to the application of nanofluids more frequently than the synthesis and determination of the physical properties of nanofluids. Thus, it is difficult to publish the updated physical properties of various nanofluids for the development of nanofluids mechanisms. Thus, it is important to publish the updated physical properties of various nanofluids systems.

In literature, various researchers have claimed the enhancement of heat transfer by 200% [30]. on performing these experiments in the laboratory, the results are very different from the reported values. These may be due to the difference in the working environment and, the difference in the precursor source. It is difficult to maintain the same working environment in all the laboratories in the world. Thus, it is important to note that the performance of nanofluids may vary in the actual application.

8.3.5 Risk to the Environment and Human Health

Nanomaterials are extensively used in the biomedical industry for gene therapy and targeted drug delivery. Medical imaging. But, still, we do not know the long-term effect of nanomaterials on human health and the environment. These nano-sized materials may be inhaled in the lungs and after significant agglomeration of these nanoparticles can lead to respiratory diseases [31].

Nanoparticles are very dangerous for the environment also. Improper treatment of nanomaterials-based waste can lead to serious environmental concerns. [32] For example, CuO is a widely used nanomaterial for various applications. But copper is also a leading contaminant in the US [33]. The copper contaminant is dangerous to marine life. Thus, open disposal of these materials into seawater can lead to serious environmental damage [34]. Similarly, copper-based nanomaterials are harmful to the earthworms and thus, the copper-based nanomaterials lead to soil pollution [35]. Various researchers have conducted a detailed study of the harmful effects of copper-based materials [33].

The research on the effect of copper sulfate and copper oxide-based nanomaterials on the human body has been done by researchers. It is found that after exposure to these nanomaterials, the poisonous effect of these nanomaterials is observed [32].

8.3.6 Scale-Up Challenges

Nanofluids are not extensively used in the industrial process. It's been already 25 years since the research was started, and still, no real-life applications of nanofluids have been reported so far. The demerits of nanofluids have remained the same for the last several years. No breakthrough research has been reported in this field. Production cost is one of the major hurdles to using nanofluids extensively for industrial applications. Apart from this, increased pumping cost, high viscosity and increased pressure drop are also important challenges for the scale-up of nanofluids. Various researchers are developing micro-channel-based modules for scale-up applications [36]. One such module is shown in Figure 8.10. As discussed in

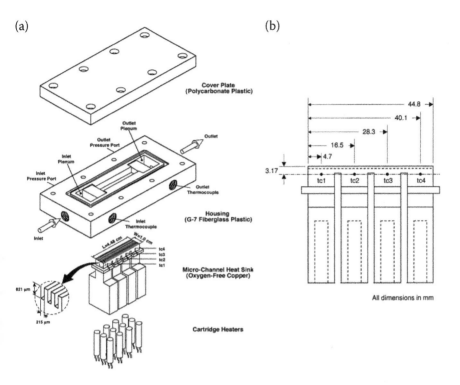

FIGURE 8.10 Micro-channel-based module [36].

Chapter 1, the stability of nanofluids is the most important parameter for the real-life application of nanofluids. The use of nanofluids without stability is not possible.

8.4 FUTURE SCOPE

Various studies and research on the thermo-physical properties of nanofluids and their application in various sectors have been reported in the last two decades. But, the widespread use of nanofluids is not yet possible [37]. Several challenges are there to use the nanofluids extensively. For example, the erosion and corrosion of pumps, heat exchangers and other mechanical parts is possible [38].

No standard procedures for the disposal of nanofluids are reported in the literature. The toxicity of used nanofluids is a major concern of the nanofluids application. The increased pumping power and pressure drop also need to be optimized for future studies of nanofluids [39]. The pressure drop of nanofluids also increases due to the increased viscosity and density of nanofluids owing to the added nanoparticles [40]. Nanomaterials are very harmful because of their small size. Thus, strategies to agglomerate the nanomaterials after their use should be developed. So, the nanofluids can be extensively used [41].

The use of the nanofluids is limited to the laboratory scale only. Thus, it is important to develop sustainable nanofluids application methods. The green synthesis routes of the nanomaterials can help to reduce the toxicity of nanofluids

[42]. The use of magnetic force also can be done to Agglomerate the nanoparticles for the nanomaterials disposal [43].

Thus, nanofluids have a good scope of research as various challenges are not yet addressed effectively and the use of nanofluids is not yet commercially produced. Thus, the development of an efficient nanofluids system is possible for a wide range of applications and a detailed study of the mechanism needs to be reported in the literature.

REFERENCES

1. L. S. Sundar, M. K. Singh, & A. C. Sousa (2014). Enhanced heat transfer and friction factor of MWCNT–Fe3O4/water hybrid nanofluids. *International Communications in Heat and Mass Transfer, 52*, 73–83.
2. A. M. Hussein (2017). Thermal performance and thermal properties of hybrid nanofluid laminar flow in a double pipe heat exchanger. *Experimental Thermal and Fluid Science, 88*, 37–45.
3. H. Yarmand, S. Gharehkhani, G. Ahmadi, S. F. S. Shirazi, S. Baradaran, E. Montazer, … & M. Dahari (2015). Graphene nanoplatelets–silver hybrid nanofluids for enhanced heat transfer. *Energy Conversion and Management, 100*, 419–428.
4. A. Akhgar, & D. Toghraie (2018). An experimental study on the stability and thermal conductivity of water-ethylene glycol/TiO2-MWCNTs hybrid nanofluid: Developing a new correlation. *Powder Technology, 338*, 806–818.
5. A. Asadi, M. Asadi, A. Rezaniakolaei, L. A. Rosendahl, & S. Wongwises (2018). An experimental and theoretical investigation on heat transfer capability of Mg (OH)₂/MWCNT-engine oil hybrid nano-lubricant adopted as a coolant and lubricant fluid. *Applied Thermal Engineering, 129*, 577–586.
6. L. Vandsburger (2009). Synthesis and covalent surface modification of carbon nanotubes for preparation of stabilized nanofluid suspensions. Department of Chemical engineering, mcQuill university, Montreal, Quebec, Cananda.
7. B. Wei, C. Zou, & X. Li (2017). Experimental investigation on stability and thermal conductivity of diathermic oil based TiO₂ nanofluids. *International Journal of Heat and Mass Transfer, 104*, 537–543.
8. S. Aberoumand & A. Jafarimoghaddam (2017). Experimental study on synthesis, stability, thermal conductivity and viscosity of Cu–engine oil nanofluid. *Journal of the Taiwan Institute of Chemical Engineers, 71*, 315–322.
9. B. Wei, C. Zou, X. Yuan, & X. Li (2017). Thermo-physical property evaluation of diathermic oil based hybrid nanofluids for heat transfer applications. *International Journal of Heat and Mass Transfer, 107*, 281–287.
10. S. KoçakSoylu, İ. Atmaca, M. Asiltürk, & A. Doğan (2019). Improving heat transfer performance of an automobile radiator using Cu and Ag-doped TiO₂ based nanofluids. *Applied Thermal Engineering, 157*, 113743.
11. S. M. Peyghambarzadeh, S. H. Hashemabadi, M. S. Jamnani, & S. M. Hoseini (2011). Improving the cooling performance of automobile radiator with Al₂O₃/water nanofluid. *Applied Thermal Engineering, 31*(10), 1833–1838.
12. G. A. Oliveira, E. M. Cardenas Contreras, & E. P. Bandarra Filho (2017). Experimental study on the heat transfer of MWCNT/water nanofluid flowing in a car radiator. *Applied Thermal Engineering, 111*, 1450–1456.
13. S. P. Jang & S. U. Choi (2006). Cooling performance of a microchannel heat sink with nanofluids. *Applied Thermal Engineering, 26*(17-18), 2457–2463.
14. A. Ijam & R. Saidur (2012). Nanofluid as a coolant for electronic devices (cooling of electronic devices). *Applied Thermal Engineering, 32*, 76–82

15. L. S. Maganti, P. Dhar, T. Sundararajan, & S. K. Das (2017). Heat spreader with parallel microchannel configurations employing nanofluids for near–active cooling of MEMS. *International Journal of Heat and Mass Transfer, 111*, 570–581. 36

16. S. S. Chatha, A. Pal, & T. Singh (2016). Performance evaluation of aluminium 6063 drilling under the influence of nanofluid minimum quantity lubrication. *Journal of Cleaner Production, 137*, 537–545.

17. M. J. Kao, C. H. Lo, T. T. Tsung, Y. Y. Wu, C. S. Jwo, & H. M. Lin (2007). Copperoxide brake nanofluid manufactured using arc-submerged nanoparticle synthesis system. *Journal of Alloys and Compounds, 434-435*, 672–674.

18. I. Zakaria, W. A. N. W. Mohamed, W. H. Azmi, A. M. I. Mamat, R. Mamat, & W. R. W. Daud (2018). Thermo-electrical performance of PEM fuel cell using Al_2O_3 nanofluids. *International Journal of Heat and Mass Transfer, 119*, 460–471

19. Investigations and theoretical determination of thermal conductivity and viscosity of TiO2-ethylene glycol nanofluid. *International Communications in Heat and Mass Transfer, 73*, 54–61. 10.1016/j.icheatmasstransfer.2016.02.004

20. K. Nishant, & S. S. Sonawane (2016). Influence of CuO and TiO_2 nanoparticles in enhancing the overall heat transfer coefficient and thermal conductivity of water and ethylene glycol based nanofluids. *Research Journal of Chemistry and Environment, 20*, 24–30

21. N. Kumar, S. S. Sonawane, & S. H. Sonawane (2018). Experimental study of thermal conductivity, heat transfer and friction factor of Al 2 O 3 based nanofluid. *International Communications in Heat and Mass Transfer, 90*, 1–10. 10.1016/j.icheatmasstransfer. 2017.10.001

22. N. Kumar, S. S. Sonawane, & S. H. Sonawane (2018). Experimental study of thermal conductivity, heat transfer and friction factor of Al2O3 based nanofluid. *International Communications in Heat and Mass Transfer, 90*, 1–10. 10.1016/j.icheatmasstransfer. 2017.10.001

23. S. Suseel Jai Krishnan, M. Momin, C. Nwaokocha, M. Sharifpur, & J. P. Meyer (2022). An empirical study on the persuasive particle size effects over the multiphysical properties of monophase MWCNT-Al_2O_3 hybridized nanofluids. *Journal of Molecular Liquids, 361*, 119668. 10.1016/j.molliq.2022.119668.

24. S. M. S. Murshed & C. A. Nieto de Castro (2014). *Nanofluids: Synthesis, properties and applications*, NOVA Science Publishers, New York, USA.

25. S. A. Angayarkanni & J. Philip (2015). Review on thermal properties of nanofluids: Recent developments. *Advances in Colloid and Interface Science, 225*, 146–176. 10. 1016/j.cis.2015.08.014.

26. C. A. Nieto de Castro & M. J. V. Lourenço (2020). Towards the correct measurement of thermal conductivity of ionic melts and nanofluids. *Energies, 13*, 99.

27. N. Sezer, M. A. Atieh, & M. Koç (2019). A comprehensive review on synthesis, stability, thermophysical properties, and characterization of nanofluids. *Powder Technology, 344*, 404–431. 10.1016/j.powtec.2018.12.016.

28. K. Franks, A. Braun, J. Charoud-Got, O. Couteau, V. Kestens, M. Lamberty, T. Linsinger, & G. Roebben (2012). *Certification of the equivalent spherical diameters of silica nanoparticles in aqueous solution—Certified Reference Material ERM®-FD304*, Publications Office of the European Union, Luxembourg.

29. J. L. Routbort, D. Singh, E. V. Timofeeva, W. Yu, & D. M. France (2011). Pumping power of nanofluids in a flowing system. *Journal of Nanoparticle Research, 13*, 931–937. 10.1007/s11051-010-0197-7.

30. M. Mehrali, E. Sadeghinezhad, M. A. Rosen, S. Tahan Latibari, M. Mehrali, H. S. C. Metselaar, & S. N. Kazi (2015). Effect of specific surface area on convective heat transfer of graphene nanoplatelet aqueous nanofluids. *Experimental Thermal and Fluid Science, 68*, 100–108. 10.1016/j.expthermflusci.2015.03.012.

31. M. Geiser, B. Rothen-Rutishauser, N. Kapp, S. Schürch, W. Kreyling, H. Schulz, M. Semmler, V. I. Hof, J. Heyder, & P. Gehr (2005). Ultrafine particles cross cellular membranes by nonphagocytic mechanisms in lungs and in cultured cells. *Environmental Health Perspectives*, *113*, 1555–1560. doi: 10.1289/ehp.8006.

32. O. Bondarenko, K. Juganson, A. Ivask, K. Kasemets, M. Mortimer, & A. Kahru (2013). Toxicity of Ag, CuO and ZnO nanoparticles to selected environmentally relevant test organisms and mammalian cells in vitro: A critical review. *Archives of Toxicology*, *87*, 1181–1200. 10.1007/s00204-013-1079-4.

33. M. J. Mashock, T. Zanon, A. D. Kappell, L. N. Petrella, E. C. Andersen, & K. R. Hristova (2016). Copper oxide nanoparticles impact several toxicological endpoints and cause neurodegeneration in Caenorhabditis elegans. *PLoS One*, *11*, e0167613.

34. A. M. Schrand, M. F. Rahman, S. M. Hussain, J. J. Schlager, D. A. Smith, & A. F. Syed (2010). Metal-based nanoparticles and their toxicity assessment. *WIREs Nanomedicine and Nanobiotechnology*, *2*, 544–568. 10.1002/wnan.103.

35. J. M. Unrine, O. V. Tsyusko, S. E. Hunyadi, J. D. Judy, & P. M. Bertsch (2010). Effects of particle size on chemical speciation and bioavailability of copper to earthworms (Eisenia fetida) exposed to copper nanoparticles. *Journal of Environmental Quality*, *39*, 1942–1953. 10.2134/jeq2009.0387.

36. J. Lee, & I. Mudawar (2007). Assessment of the effectiveness of nanofluids for single-phase and two-phase heat transfer in micro-channels. *International Journal of Heat and Mass Transfer*, *50*, 452–463. 10.1016/j.ijheatmasstransfer.2006.08.001.

37. P. Thakur, N. Kumar, & S. S. Sonawane (2021). Enhancement of pool boiling performance using MWCNT based nanofluids: A sustainable method for the wastewater and incinerator heat recovery. *Sustainable Energy Technologies and Assessments*, *45*, 101115. 10.1016/j.seta.2021.101115

38. S. Sonawane, P. Thakur, S. H. Sonawane, & B. A. Bhanvase (2021). Nanomaterials for membrane synthesis: Introduction, mechanism, and challenges for wastewater treatment. In: *Handbook of nanomaterials for wastewater treatment*. Elsevier, pp. 537–553

39. M. Malika, R. Bhad, & S. S. Sonawane (2021). ANSYS simulation study of a low volume fraction CuO–ZnO/water hybrid nanofluid in a shell and tube heat exchanger. *Journal of the Indian Chemical Society*, *98*, 100200. 10.1016/j.jics.2021.100200

40. M. Malika & S. S. Sonawane (2021) Review on CNT based hybrid nanofluids performance in the nano lubricant application. *Journal of Indian Association for Environmental Management*, *41*, 1–16

41. M. Malika & S. S. Sonawane (2021) A comprehensive review on the effect of various ultrasonication parameters on the stability of nanofluid. *Journal of Indian Association for Environmental Management*, *41*, 19–25

42. U. S. Shenoy & A. N. Shetty (2018). A simple single-step approach towards synthesis of nanofluids containing cuboctahedral cuprous oxide particles using glucose reduction. *Frontiers of Materials Science*, *12*, 74–82. 10.1007/s11706-018-0411-6

43. M. Malika, P. G. Jhadav, & V. R. Parate et al. (2022). Synthesis of magnetite nanoparticle from potato peel extract: its nanofluid applications and life cycle analysis. *Chemical Papers*, 10.1007/s11696-022-02538-w

Index

For Product Safety Concerns and Information please contact our EU
representative GPSR@taylorandfrancis.com
Taylor & Francis Verlag GmbH, Kaufingerstraße 24, 80331 München, Germany

www.ingramcontent.com/pod-product-compliance
Ingram Content Group UK Ltd.
Pitfield, Milton Keynes, MK11 3LW, UK
UKHW021109180425
457613UK00001B/10